ROBOT-PROOF

ROBOT-PROOF

Higher Education in the Age of Artificial Intelligence

JOSEPH E. AOUN

The MIT Press
Cambridge, Massachusetts
London, England

© 2017 Massachusetts Institute of Technology

All rights reserved. No part of this book may be reproduced in any form by any electronic or mechanical means (including photocopying, recording, or information storage) without permission in writing from the publisher.

This book was set in Scala Pro by Toppan Best-set Premedia Limited. Printed and bound in the United States of America.

Library of Congress Cataloging-in-Publication Data is available.

ISBN: 978-0-262-03728-0

10   9   8   7   6   5   4   3   2   1

CONTENTS

ACKNOWLEDGMENTS

A great many people at Northeastern University have contributed to the ideas and concepts discussed in this book. Foremost, I thank J. D. LaRock and Andrew Rimas, without whom this project would not have been completed.

I also thank my colleagues Michael Armini, James Bean, James Hackney, Diane MacGillivray, Philomena Mantella, Ralph Martin, and Thomas Nedell. Our work together has informed much that is written here. Susan Ambrose and Uta Poiger also provided invaluable insights, particularly regarding experiential learning, the science of learning, and the "experiential liberal arts."

I have drawn liberally from Northeastern's academic plan, "Northeastern 2025," for many of the discussions herein, including about the new learning model, "humanics." I thank my faculty colleagues, staff colleagues, and students for contributing to this deep and forward-looking document.

I also thank Northeastern's board of trustees, including trustee leaders Neal Finnegan, Sy Sternberg, Henry Nasella, and Rich D'Amore, who have supported our efforts to bring many of the ideas and themes discussed here into practice at the university.

I am continually grateful for the support of former colleagues and mentors who have helped to shape my thinking about higher education and the world, including Lloyd Armstrong and Vartan Gregorian.

This book also benefits from insights revealed over the course of interviews and conversations with students, scholars, and business leaders beyond those quoted in the pages that follow. I thank my Northeastern colleagues Chris Gallagher, Dan Gregory, Marc Meyer, Dennis Shaughnessy, Maria Stein, Alan Stone, Cigdem Talgar, and Michelle Zaff for their reflections.

Finally, I owe everything to the love and support of my wife, Zeina, and my sons, Adrian and Karim.

INTRODUCTION

Thousands of years ago, the agricultural revolution led our foraging ancestors to take up the scythe and plough. Hundreds of years ago, the Industrial Revolution pushed farmers out of fields and into factories. Just tens of years ago, the technology revolution ushered many people off the shop floor and into the desk chair and office cube.

Today, we are living through yet another revolution in the way that human beings work for their livelihoods—and once again, this revolution is leaving old certainties scrapped and smoldering on the ash heap of history. Once again, it is being powered by new technologies. But instead of the domesticated grain seed, the cotton gin, or the steam engine, the engine of this revolution is digital and robotic.

We live in a time of technological marvels. Computers continue to speed up while the price of processing power continues to plummet, doubling and redoubling the capabilities of machines. This is driving the advance of machine learning—the ability of computers to learn from data instead of from explicit programming—and the push for artificial intelligence. As economists Erik Brynjolfsson and Andrew McAfee note in their book *The Second Machine Age: Work, Progress, and Prosperity in a Time of Brilliant Technologies*, we have recently hit an inflection point in which our machines have reached their "full force" to transform the world as comprehensively as James Watt's engine transformed an economy that once trundled along on ox carts.[1] Labor experts are increasingly and justifiably worried that computers are becoming so adept at human capabilities that soon there will be no need for any human input at all.[2]

The evidence for this inflection point is everywhere. Driverless cars are now traversing the streets of Pittsburgh, Pennsylvania, and other cities. New robots can climb stairs and open doors with ease. An advanced computer trounced the human grandmaster of the intricate Chinese strategy game Go. Moreover, it is not only the processing power of machines that has skyrocketed exponentially but also the power of their connectivity, their sensors, their GPS systems, and their gyroscopes. Today, we are giving computers not only artificial intelligence but, in effect, artificial eyes, ears, hands, and feet.

Consequently, these capacities are enabling computers to step into roles—and jobs—once held exclusively by members of our species. Robots now analyze stocks, write in deft and informative prose, and interact with customers.[3] Semi-autonomous machines may soon join soldiers on the battlefield.[4] In China, "co-bots"—machines that can work in factories safely alongside human beings—are upending that country's vaunted manufacturing sector, allowing fewer laborers to be vastly more productive. In 2015, sales of industrial robots around the world increased by 12 percent over the previous year, rising to nearly a quarter of a million units.[5]

At the same time, Big Data is revolutionizing everything from social science to business, with organizations amassing information in proportions that flirt with the infinite. Algorithms mine bottomless troves of data and then apply the information to new functions, essentially teaching themselves. Machine learning now powers everything from our spam filters to our Amazon shopping lists and dating apps, telling us what to watch, what to buy, and whom to love.[6] "Deep learning" systems, in which artificial neural networks identify patterns, can now look at an image and recognize a chair or the face of a human individual or teach themselves how to play a video game without ever reading the instructions.[7]

In many ways, these new technologies are an astonishing boon for humanity, giving us the power to mitigate poverty, hunger, and disease. For example, Stanley S. Litow, vice president of corporate citizenship and corporate affairs at IBM, is overseeing an initiative between Memorial Sloan Kettering Hospital in New York City and Watson, the computer that famously beat the human champions of the television game show *Jeopardy!* A doctor who had watched the show approached IBM with the idea to collaborate. Thus, Watson was reborn as an oncology adviser. Computer scientists at IBM embedded it with information from the hospital's clinical trials ("not just some, *all* of them," said Litow)[8] and trained it through data analytics to respond to oncologists' questions.

"So it proceeds as if talking to a potential patient," said Litow. "On a mobile device I can say, 'She has the following characteristics. Do we have any information on clinical trials that would help me figure out whether this is the problem or that is the problem?'" Watson then analyzes the data and responds to the oncologist's question in normal English. "There's a lot of clinical trial information, but a lot of doctors don't have access to it," said Litow. "It is actually helping some of the best oncologists in the United States make a better, faster diagnosis and move toward a treatment plan quickly. In treating cancer, that's critical."

Watson's next challenge is to improve teaching in the New York City public school system, advising educators on effective teaching practices by using the same data analytics and communication techniques it is deploying with such success at Sloan Kettering. Technologies like Watson are helping people save lives, teach fractions, and—in their less sophisticated iterations—find the nearest parking space. They are helping people work better.

Or they are, for the moment. Automation long has been considered a threat to low-skilled labor, but increasingly, any predictable work—including many jobs considered "knowledge economy" jobs—are now within the purview of machines.[9] This includes many high-skill functions, such as interpreting medical images, doing legal research, and analyzing data.

As advanced machines and computers become more and more proficient at picking investments, diagnosing disease symptoms, and conversing in natural English, it is difficult not to wonder what the limits to their capabilities are. This is why many observers believe that technology's potential to disrupt our economy—and our civilization—is unprecedented.

Over the past few years, my conversations with students entering the workforce and the business leaders who hire them have revealed something important: to stay relevant in this new economic reality, higher education needs a dramatic realignment. Instead of educating college students for jobs that are about to disappear under the rising tide of technology, twenty-first-century universities should

liberate them from outdated career models and give them ownership of their own futures. They should equip them with the literacies and skills they need to thrive in this new economy defined by
technology, as well as continue providing them with access to the
learning they need to face the challenges of life in a diverse, global
environment. Higher education needs a new model and a new ori-
entation away from its dual focus on undergraduate and graduate
students. Universities must broaden their reach to become engines
for lifelong learning.

There is a great deal of evidence that we need such an educational
shift. An oft-quoted 2013 study from Oxford University found that
nearly half of U.S. jobs are at risk of automation within the next
twenty years.[10] In many cases, that prediction seems too leisurely.
For example, new robotic algorithmic trading platforms are now
tearing through the financial industry, with some estimates holding
that software will replace between one-third and one-half of all
finance jobs in the next decade.[11] A 2015 McKinsey report found
that solely by using existing technologies, 45 percent of the work
that human beings are paid to do could be automated, obviating
the need to pay human employees more than $2 trillion in annual
wages in the United States.[12]

This is not the first time we have faced a scenario like this. In past
industrial revolutions, the ploughmen and weavers who fell prey to
tractors and spinning jennies had to withstand a difficult economic
and professional transition. However, with retraining, they could
reasonably have expected to find jobs on the new factory floors.
Likewise, as the Information Age wiped out large swaths of manufacturing, many people were able to acquire education and training
to obtain work in higher-skilled manufacturing, the service sector,

or the office park. Looking ahead, education will remain the ladder by which people ascend to higher economic rungs, even as the jobs landscape grows more complex. And it undoubtedly is getting knottier. One of the reasons for this is that the worldwide supply of labor continues to rise while the net number of high-paying, high-productivity jobs appears to be on the decline.[13] To employ more and more people, we will need to create more and more jobs. It is not clear where we will find them.

Certainly, the emergence of new industries—such as those created in the tech sector—will have to step up if they are going fill this gap. According to the U.S. Bureau of Labor Statistics, the computer and information technology professions are projected to account for a total of 4.4 million jobs by 2024.[14] In the same period, the labor force, aged sixteen and older, is expected to reach 163.7 million. Adding to the disjoint is the remarkable labor efficiency of tech companies. For instance, Google, the standard bearer for the new economy, had 61,814 full-time employees in 2015. At its peak in 1979, in contrast, General Motors counted 600,000 employees on its payroll.[15] To address the deficit, we'll need creative solutions.

Apart from automation, many other factors are stirring the economic pot. Globalization is the most apparent, but environmental unsustainability, demographic change, inequality, and political uncertainty are all having their effects on how we occupy our time, how we earn our daily bread, and how we find fulfillment. Old verities are melting fast. The remedies are not obvious.

Some observers have been encouraged by the growth of the "gig economy," in which people perform freelance tasks, such as driving a car for Uber, moving furniture through TaskRabbit, or typing text for Amazon Mechanical Turk. But earnings through these plat-forms are limited. Since 2014, the number of people who earn 50 percent or more of their income from "gig" platforms has actually fallen.[16] In general, these platforms give people a boost to earnings and help to pay the monthly bills. But as an economic engine, they have not emerged as substitutes for full-time jobs.

Of the new full-time jobs that are appearing, many are so-called hybrid jobs that require technological expertise in programming or data analysis alongside broader skills.[17] Fifty years ago, no one could have imagined that user-experience designer would be a legitimate profession, but here we are. Clearly, work is changing. All these factors create a complex and unexplored terrain for job seekers, begging some important questions: How should we be preparing people for this fast-changing world? How should edu-cation be used to help people in the professional and economic spheres?

As a university president, this is no small question for me. As a matter of fact, the university I lead, Northeastern, is explicitly concerned with the connections between education and work. As a pioneer in experiential learning, grounded in the co-op model of higher education, Northeastern's mission has always been to prepare students for fulfilling—and successful—roles in the pro-fessional world. But lately, as I have observed my students try to puzzle out their career paths, listened to what employers say they are looking for in new employees, and take stock of what I read and hear every day about technology's impact on the world of pro-fessional work, I have come to realize that the existing model of higher education has yet to adapt to the seismic shifts rattling the foundations of the global economy.

I believe that college should shape students into professionals but also creators. Creation will be at the base of economic activity and also much of what human beings do in the future. Intelligent machines may liberate millions from routine labor, but there will remain a great deal of work for us to accomplish. Great undertakings like curing disease, healing the environment, and ending poverty will demand all the human talent that the world can muster. Machines will help us explore the universe, but human beings will face the consequences of discovery. Human beings will still read books penned by human authors and be moved by songs and artworks born of human imagination. Human beings will still undertake ethical acts of selflessness or courage and choose to act for the betterment of our world and our species. Human beings will also care for our infants, give comfort to the infirm, cook our favorite dishes, craft our wines, and play our games. There is much for all of us to do.

To that end, this book offers an updated model of higher education—one that will develop and empower a new generation of creators, women and men who can employ all the technological wonders of our age to thrive in an economy and society transformed by intelligent machines. It also envisions a higher education that continues to deliver the fruits of learning to students long after they have begun their working careers, assisting them throughout their lives. In some ways, it may seem like a roadmap for taking higher education in a new direction. However, it does not offer a departure as much as a continuity with the centuries-old purpose of colleges and universities—to equip students for the rigors of an active life within the world as it exists today and will exist in the future. Education has always served the needs of society. It must do so now, more than ever. That is because higher education is the usher of progress and change. And change is the defining force of our time.

A UNIQUELY HUMAN EDUCATION

Education is its own reward, equipping us with the mental furniture to live a rich, considered existence. However, for most people in an advanced society and economy such as ours, it also is a prerequisite for white-collar employment. Without a college degree, typical employees will struggle to climb the economic ladder and may well find themselves slipping down the rungs.

When the economy changes, so must education. It has happened before. We educate people in the subjects that society deems valuable. As such, in the eighteenth century, colonial colleges taught classics, logic, and rhetoric to cadres of future lawyers and clergymen. In the nineteenth century, scientific and agricultural colleges rose to meet the demands of an industrializing world of steam and steel. In the twentieth century, we saw the ascent of professional degrees suited for office work in the corporate economy.

Today, the colonial age and the industrial age exist only in history books, and even the office age may be fast receding into memory. We live in the digital age, and students face a digital future in which robots, software, and machines powered by artificial intelligence perform an increasing share of the work humans do now. Employment will less often involve the routine application of facts, so education should follow suit. To ensure that graduates are "robot-proof" in the workplace, institutions of higher learning will have to rebalance their curricula.

A robot-proof model of higher education is not concerned solely with topping up students' minds with high-octane facts. Rather, it refits their mental engines, calibrating them with a creative mindset and the mental elasticity to invent, discover, or otherwise produce something society deems valuable. This could be anything at all—a scientific proof, a hip-hop recording, a new workout regimen, a web comic, a cure for cancer. Whatever the creation, it must in some manner be original enough to evade the label of "routine" and hence the threat of automation. Instead of training laborers, a robot-proof education trains creators.

The field of robotics is yielding the most advanced generation of machines in history, so we need a disciplinary field that can do the same for human beings. In the pages that follow, I lay out a framework for a new discipline—"humanics"—the goal of which is to nurture our species' unique traits of creativity and flexibility. It builds on our innate strengths and prepares students to compete in a labor market in which brilliant machines work alongside human professionals. And much as today's law students learn both a specific body of knowledge and a legal mindset, tomorrow's humanics students must master specific content as well as practice uniquely human cognitive capacities.

In the chapters ahead, I describe both the architecture and the inner workings of humanics, but here I begin by explaining its twofold nature. The first side, its content, takes shape in what I call the *new literacies*. In the past, literacy in reading, writing, and mathematics formed the baseline for participation in society, while even educated professionals did not need any technical proficiencies beyond knowing how to click and drag through a suite of office programs. That is no longer sufficient. In the future, graduates will need to build on the old literacies by adding three more—*data literacy, technological literacy,* and *human literacy.* This is because people can no longer thrive in a digitized world using merely analog tools. They will be living and working in a constant stream of big data, connectivity, and instant information flowing from every click and touch of their devices. Therefore, they need data literacy to read, analyze, and use these ever-rising tides of information. Technological literacy gives them a grounding in coding and engineering principles, so they know how their machines tick. Lastly, human literacy teaches them humanities, communication, and design, allowing them to function in the human milieu.

As noted earlier, knowledge alone is not sufficient for the work of tomorrow. The second side of humanics, therefore, is not a set of content areas but rather a set of *cognitive capacities.* These are higher-order mental skills—mindsets and ways of thinking about the world. The first is *systems thinking,* the ability to view an enterprise, machine, or subject holistically, making connections between its different functions in an integrative way. The second is *entrepreneurship,* which applies the creative mindset to the economic and often social sphere. The third is *cultural agility,* which teaches students how to operate deftly in varied global environments and to see situations through different, even conflicting, cultural lenses. The fourth capacity is that old chestnut of liberal arts programs, *critical thinking,* which instills the habit of disciplined, rational analysis and judgment.

Together, the new literacies and the cognitive capacities integrate to help students rise above the computing power of brilliant machines by engendering creativity. In doing so, they enable them to collaborate with other people and machines while accentuating the strengths of both. Humanics can, in short, be a powerful toolset for humanity.

This book also explores how people grasp these tools. To acquire the cognitive capacities at a high level, students must do more than read about them in the classroom or apply them in case studies or classroom simulations. To cement them in their minds, they need to experience them in the intensity and chaos of real work environments such as co-ops and internships. Just as experiential learning is how toddlers puzzle out the secrets of speech and ambulation, how Montessori students learn to read and count, and how athletes and musicians perfect their jump shots or arpeggios, it also is how college students learn to think differently. This makes it the ideal delivery system for humanics.

A new model of higher education must, however, account for the fact that learning does not end with the receipt of a bachelor's diploma. As machines continue to surpass their old boundaries, human beings must also continue to hone their mental capacities, skills, and technological knowledge. People rarely stay in the same career track they choose when they graduate, so they need the support of lifelong learning. Universities can deliver this by going where these learners are. This means a fundamental shift in our delivery of education but also in our idea of its timing. It no longer is sufficient for universities to focus solely on isolated years of study for undergraduate and graduate students. Higher education must broaden its view of whom to serve and when. It must serve everyone, no matter their stage in life.

By 2025, our planet will count eight billion human inhabitants, all of them with human ambition, intelligence, and potential.[18] Our planet will be more connected and more competitive than the one we know today. Given the pace of technology's advance, we can predict that computers, robots, and artificial intelligence will be even more intricately intertwined into the fabric of our personal and professional lives. Many of the jobs that exist now will have vanished. Others that will pay handsomely have yet to be invented. The only real certainty is that the world will be different—and with changes come challenges as well as opportunities. In many cases, they are one and the same.

Education is what sets them apart.

FEARS OF A ROBOTIC FUTURE

The upshot is simply a question of time, but that the time will come when the machines will hold the real supremacy over the world and its inhabitants is what no person of a truly philosophic mind can for a moment question.

—Samuel Butler, "Darwin among the Machines" (1863)

In 2015, Chapman University published the results of a survey ranking the U.S. public's worst fears. "Man-made disasters" such as terrorism and nuclear attacks stood at the top of the list of popular horrors. But in close second place—even more terrifying than crime, earthquakes, and public speaking—was fear of technology. In fact, technology appears to frighten many of us more than the absolute unknown. According to the survey, Americans fear robots replacing people in the workforce more than they fear death—and by a full seven percentage points.[1]

But it is not paranoia if they really are out to get you. Machines have been replacing human labor ever since a piece of flint proved to be sharper than a fingernail. The history of workplace obsolescence is almost as old as the history of work. As technologies increase our capacity for labor, the nature of labor changes. The question is whether the evolution of work in the twenty-first century is qualitatively different from the evolution of work in the twentieth, the nineteenth, or indeed, the tenth century BCE.

ELEMENTS AND WORK

In physics, work is done when a force is applied to an object, moving it in a direction. This expends energy. In biology, all organisms expend energy to obtain nourishment and to continue the process of living, expending, and feeding.

Throughout history, human beings have spent most of their existence expending energy on work to obtain food. But unlike many other organisms, we have invented ways to amplify that energy by harnessing forces far greater than those available to us in our teeth and musculature. Perhaps as early as a million years ago, our ancestors tamed the element of fire.[2] Controlled fire was among the greatest of all work innovations. By cooking food, our ancestors were able to spend less energy in digestion, allowing us to eat useful plants like wheat and rice, destroying bacteria that taxes our bodies, and reducing the work we spend in chewing and processing. This freed us to expend more energy on evolving our enormous brains.[3]

Much more recently, human beings tamed plants and livestock, vastly increasing the amount of energy we could consume and, in the case of draft animals, deploy for work. We also harnessed the element of air through the invention of sails and windmills. But when we tamed the power of steam, we found a truly reliable elemental force.

The Industrial Revolution began with the realization that heat could cause movement, which performs work. By boiling water, you could move a piston, which could in turn move anything an eighteenth-century engineer might attach to it. Beginning with water pumps in mineshafts, machines started to do work that previously had been the provenance of strong limbs and strained backs. Within decades, this process of industrialization transformed almost every aspect of human society.

The world turned mechanical, tapping into coal and then oil to generate seemingly unlimited amounts of energy. Factories and mills roared to life, railroads chugged across the countryside, gas-lit cities mushroomed with brick and iron, their populations teeming on the fruits of the seed drill and the Dutch plough. Not since the dawn of agriculture had humanity experienced such incontrovertible change.

Yet only at the end of the nineteenth century did the full power of the technological revolution come into force. This was brought on with the taming of electricity by scientific discoverers such as Michael Faraday and inventors such as Thomas Edison and Joseph Swan. In 1881, Swan used his incandescent bulbs to illuminate London's Savoy Theater, and in a few short years, electricity freed humanity from billions of years of nocturnal darkness. With the development of high-voltage alternating current, engineers were able to power assembly lines and mass production, again amplifying our capacity for physical work and revolutionizing the way we live.

Fire, steam, and electricity have been the three elemental forces that amplified humanity's energy to perform work. Then, in the middle of the twentieth century, a new force appeared with the potential to be equally transformative. Information—the ones and zeroes that fuel our digital machines—is proving just as titanic a force for change as any of its predecessors. Indeed, because digital power amplifies our capacity for mental work, it may be more transformative than any force since an ancient hominid first learned to strike a fire.

As in physics, when a body performs work, applying force to move an object in a particular direction, the object simultaneously exerts a force equal in magnitude and opposite in direction on the first body. In other words, for every action, there is an equivalent opposition. And that surely has been the case as the force of technology has acted on human society.

THE RETURN OF ROBIN HOOD

On February 27, 1812, the young George Gordon Byron, the sixth Baron Byron, stood before the House of Lords to deliver his first address. Although a few days later, Lord Byron published the first two cantos of *Childe Harold's Pilgrimage* and became an instant celebrity, when he took the podium in Parliament, he was still a relatively unknown scribbler of satire and amorous verse, not yet the "mad, bad and dangerous to know" character who shocked drawing room society. Even so, his first speech as a politician was about a scandalous event.

As well as being the legendary home of Robin Hood, the city of Nottingham long had been a center for the manufacture of hosiery. However, by 1812, technological innovations were upending the stocking business as the town's factory owners introduced steam-powered mechanical frames, replacing the labor of skilled artisans. These artisans possessed highly developed—but very particular—skill sets that the marketplace no longer needed. Consequently, desperate to save their livelihoods, the indigent laborers formed secret societies under the banner of an invented character, Ned Ludd—a Robin Hood figure updated for the industrial age. Calling themselves "Luddites," in November 1811, they broke into the hosiery factories and smashed the owners' new machines. The uprising soon spread to surrounding communities, forcing the government to call in the military. At one point, more British soldiers were battling the Luddites than were deployed against the French in the Iberian peninsula.[4]

Lord Byron owned land in Nottinghamshire and had witnessed the violence and disruption firsthand. So when the House of Lords sat to debate whether to make frame-breaking a capital offense, he spoke passionately in defense of the Luddites, arguing that the rioters' "own means of subsistence were cut off, all other employment preoccupied; and their excesses, however to be deplored and condemned, can hardly be subject to surprise."[5] In other words, if machines took the weavers' work, they hardly could be blamed for wanting them smashed.

Byron's eloquence notwithstanding, the act passed. Several days later on March 2, the *London Morning Chronicle* published an anonymous poem titled "Ode to the Framers of the Frame Bill," although its authorship by Byron was not hard to figure out. Among its more scathing verses:

Men are more easily made than machinery—
Stockings fetch better prices than lives—
Gibbets on Sherwood will heighten the scenery,
Showing how Commerce, how Liberty thrives![6]

For two hundred years, the Luddites have been a symbol of resistance to technological displacement—and over those two hundred years, there has been a great deal of displacement to symbolize. The invention of the tractor took manual laborers off the land and into factories. The development of automated processes in factories took employees off the assembly lines and into the corporate office park. Karl Marx warned of the effects of automation on the proletariat, and John Maynard Keynes believed that machines would cause "technological unemployment."[7]

By the middle of the twentieth century, people's fear of displacement by machines did not apply just to factory laborers. Even as the postwar economy of the 1940s and 1950s saw a huge shift away from manual work to clerical and professional work, as early as 1964, President Lyndon B. Johnson received an open letter from a group of prominent academics warning of technology's potential to undermine the value of *all* human labor.[8]

When farm laborers left their ploughs for city jobs, they needed new skills to function effectively in industrial workplaces. Generations later, when they abandoned their lathes and welding irons for typewriters and dictation machines, their descendants needed to upskill once more. As a matter of fact, when grappling with technological and social changes, people have always responded by improving their education.

THE ENGINE OF PROGRESS

At its best, higher education does not mirror society from a distance. It is not apart from it but runs like a thread through its fabric, conforming to its patterns. Since the emergence of universities in medieval Europe, their chief purpose has always been to equip students for the economic and professional roles of the day. Before Nicolaus Copernicus and Isaac Newton, universities were largely concerned with training ministers, lawyers, and teachers. The economies of medieval Italy, England, and Spain needed literate individuals to conduct affairs of the soul and the state, to record agreements, and to administrate property and institutions. So that is what the colleges of Bologna, Oxford, and Salamanca produced.

In the 1850s, the United States was mostly rural, agrarian, and unlearned. There was no need for more higher education than what was offered by a handful of colonial colleges dealing in what Cardinal John Henry Newman, the theologian and nineteenth-century intellectual, called "liberal knowledge," the purpose of which was to prepare men "to fill any post with credit, and to master any subject with facility." Moreover, Newman believed that the most valuable education would cultivate a man who "is able to converse ... is able to listen ... can ask a question pertinently ... [is] yet never in the way ... [and has] a sure tact which enables him to trifle with gracefulness and to be serious with effect."[9] In short, the colleges of that age largely prepared men to become gentlemen who would thrive in a technologically undemanding but culturally rich economy and society.

Yet even as Newman wrote, that world was changing. Just as the Industrial Revolution remade society in the image of its machines and companies, it also remade higher education. Less than a hundred years after James Watt fired his engine, the U.S. Congress passed the Morrill Act of 1862, giving public land to endow universities that would train a new generation of technological masters. Their goal was "without excluding other scientific and classical studies and including military tactic, to teach such branches of learning as are related to agriculture and the mechanic arts," the new technologies of the day. They accomplished this by modeling the United States' new colleges and universities on the great German research universities that had emerged after the Napoleonic wars.[10]

The new land grant and research universities evolved past the old liberal arts curricula to focus on nonclassical languages, the newly emerging field of the social sciences, and scientific and technological discovery. Building on scientific principles, new branches of ingenuity shot forth from the laboratories and lecture halls. Disciplines like economics, biology, and engineering coalesced around growing faculties. Instead of teaching knowledge dating back to the Greeks and Romans, higher education began to devote its energies to the active creation of new knowledge. Instead of simply cultivating the individual, universities took on the work of cultivating economic and social progress.[11]

Progress required the individual's participation, so the individual needed the appropriate schooling. As early as the 1830s, educators in the United States were looking overseas to Prussia for ideas on how to formalize a system of education for the nation's children. Reformers such as Horace Mann advocated a form of schooling that was free, universal, and nonsectarian and that would teach children how to be good citizens and participants in a modern republic.

In 1848, Mann introduced this Prussian model to Massachusetts, establishing the basis for much of the K–12 system that persists to this day. And although it now is fashionable to criticize it as a "factory model" designed to batch-process masses of students to enter roles in an industrial economy—cutting cogs to fit the machine—it successfully educated generations of young Americans for the demands of their times. Until the 1940s, that meant joining a massive migration to urban centers and the rapid mechanization of work.

On June 22, 1944, with American troops still battling through the hedgerows behind Omaha Beach, U.S. higher education undertook its next pivotal transformation. Anticipating the return of millions of veterans into the fold of civic life and the need to integrate them into the economy, Congress passed the Servicemen's Readjustment Act—better known as the G.I. Bill—one of the benefits of which was provision of tuition and living expenses for college attendance.

Not since the land-grant movement of the 1860s had there been such a dramatic widening of access to higher education. The returning veterans flooded in, soon becoming more than 50 percent of the country's college population. By 1956, the G.I. Bill had helped more than 2.2 million Americans attend college.[12] To accommodate these huge numbers of new students, universities needed to expand radically, and they did so through a huge investment to expand state's public higher education systems, including creating a new type of school—the community college.

Accommodating these new students also meant shifting what colleges and universities taught. In April 1947, *Life* magazine featured a cover story on the influx of student veterans, about whom the magazine wrote: "The veteran student is poor and hard-working. He has been around enough to make subjects like geography tough to teach. He wants a fast, business-like education and is doing his best to see that he gets it."[13] In other words, colleges' new customers were taking stock of the economy and society around them and demanding something different from higher education institutions than what they provided before.

World War II did not only transform the demographics and culture within university classrooms. It also changed the way they operated their laboratories and institutes. When the Germans invaded France in 1940, Vannevar Bush, then head of the Carnegie Institution, approached President Franklin D. Roosevelt with a brief, one-page proposal for a National Defense Research Committee. The idea was to coordinate research between military officials and academics, effectively adding the weight of universities to the war effort by throwing them wholeheartedly into scientific and technological discovery. It also opened the sluices for a flood of federal money. Federal dollars launched wartime laboratories at universities such as the Massachusetts Institute of Technology, the University of California at Berkeley, and the University of Chicago.[14]

The collaboration between the military and academy reached its most famous apotheosis in the Manhattan Project, but the cessation of hostilities did not mean the end of research. Throughout the Cold War and beyond, government funding continued to flow to universities, buoying the creation of scientific and technological knowledge. In the 1960s, federal funding accounted for 73 percent of university research and development budgets. Today, it has dropped to about 60 percent but still amounts to approximately $30 billion annually.[15] To say that this marriage of government and academy proved fruitful is a gross understatement: it has given the world everything from the digital computer to the jet airliner to the commercial polio vaccine. More than ever before, universities became loci of creativity.

Since the war, then, higher education has acted as a force for progress in two ways. Through knowledge creation, universities are themselves the engines of technological progress. When the twentieth century began, 86 percent of the world's 1.6 billion human beings lived on farms, living and dying by firelight.[16] When the century ended, about half of the world's six billion people lived in cities humming with electricity.[17] Human beings had walked on the moon, split the atom, and leveled whole cities using knowledge discovered by university scientists. They treated their ailments and extended their lifespans through technologies invented in university research labs. Their computers communicated through networks developed by universities spending government dollars.

Just as powerfully, higher education is also a force for individual progress. Universities deliver the skills people need to advance their careers as technology and the economy pushes forward. This happened when the G.I. Bill extended the benefits of higher education to millions of Americans, equipping them for the knowledge economy as the first tremors of globalization and automation began to tilt work from the factories to the service sector.

It is not a coincidence that the emergence of widespread higher education and the growth of the middle class coincided in the latter half of the twentieth century. As companies grew more complex, people needed more training to fill the roles of accountant, lawyer, and manager. There was a clear link between a university degree and an employee's ascent up the corporate ladder. Indeed, the connection between the two remains immutable. The so-called wage premium for having a college degree had risen steadily since the 1960s, eventually reaching a median of about 80 percent higher hourly earnings over people with solely high school diplomas.[18]

A WRENCH IN THE ENGINE

For thousands of years, human beings worked the land. Two hundred years ago, machines displaced farm laborers because they were physically stronger and faster at the grueling tasks demanded by agriculture. Some of these farmers found better lives working in the industrial economy, applying themselves to rote work that required some, but not much, education. In turn, their descendants eventually began to surrender those factory jobs to machines that were more efficient at routine tasks, requiring industrial employees to educate themselves further in order to rise to better positions in corporate offices. Finally, in the late twentieth century, computers began to perform routine cognitive tasks with an efficiency that no human being could match, invading the accountancy office, the call center, and the secretarial pool.

Cycles of automation and disruption generally have led to elevated living standards and economic growth as people found jobs performing work that machines could not. But as machines have sped up, so have the cycles. As computers and advanced machines take the next leap forward and attain high levels of cognition, they are poised to replace professionals who make decisions based on information: in other words, they are poised to replace thinkers. We now have machines that write news articles, translate foreign languages, and interact pleasantly with customers. We have machines that edit the genomes of the life we know and scan the universe for the life we do not. Machines build our automobiles and will soon drive them. On Wall Street, they are ousting financial analysts by the hundreds, with some observers estimating that between one-third and one-half of all finance employees will be replaced by software within the next ten years.[19] In a few years, they may stand in for human surgeons in performing operations like appendectomies.[20]

Scott Semel, the chief executive officer of "cloud-based Content Collaboration Network" Intralinks, has said that machines are much better than legal associates at scanning and summarizing large numbers of leases or licensing agreements. "The A.I. just does the same thing over and over and over and over again," said Semel. "People get tired. Two different people could read the same contract. Somebody could finish half of it and go home, stay out too late, come back hungover. There are lots of variables around that. And that kind of work, which is summarizing, distilling lots of data into buckets, it's something that machines can do well."[21]

Shelves of recent books have delved into the economic implications of the emergence of intelligent machines. Klaus Schwab, founder of the World Economic Forum, released a title on the subject prior to the group's 2016 annual meeting, where automation was, not coincidentally, at the top of the agenda.[22] Magazines and news sites do a brisk business in headlines forecasting the end of work, the new machine age, and the debasement of the value of human labor. A sizeable share of this pontification and analysis has a Cassandra timbre, warning of woeful consequences unless we undertake drastic and politically thorny steps like instituting a universal basic income. Most of them conclude that employment will grow scarcer.

But not everyone views AI as a threat to human labor. Colin Angle, CEO and cofounder of robot manufacturer iRobot Corporation, has noted, "When computers came in, it was going to revolutionize how people did business. It didn't. It certainly helped people be more efficient, but it didn't eliminate jobs so much as create opportunities to do more."[23]

Angle's company produces the popular Roomba robotic vacuum cleaner, which is essentially a time-saving device, a direct descendant of the dishwasher. He believes the new technologies emerging today—artificial intelligence, advanced machines, and supercomputers—are inherently similar. We should not make the mistake of "thinking the world is a closed, zero-sum system and new opportunities won't be created by these technologies," said Angle. "I think history has proven that's a very inaccurate view of how the world works."

Historically, he is right. There is a Malthusian gloom to the idea that the labor market is a singular clump, of which the lion's share is eaten by robots and the scraps are divvyed up by ever-hungrier humans. But over the centuries, Malthus's idea of abundance leading to overpopulation and collapse has proved spectacularly incorrect. Since the Industrial Revolution first put machines in the roles of human beings, we always have found new industries and new frontiers for our talents. Instead of fighting for the same jobs, we have invented new ones.

At the same time, it is clear that the current digital revolution is different from previous technological leaps because machines now seem to have no limit to their potential processing power—no limit to their intelligence. In any predictable task, computers have humans at a cognitive disadvantage. And because software is cheap to copy, any digital advance can be instantly replicated throughout the world. As the technology writer Martin Ford has observed, "Imagine the impact of a large corporation being able to train a single employee and then clone him into an army of workers, all of whom instantly possess his knowledge and experience but, from that point on, are also capable of continuing to learn and adapt to new situations."[24] If this is, indeed, the technological near future—and there are many reasons to believe that it is—we could be living in a time in which paid human labor becomes an anomaly.

Nor are smart machines the only source of pressure on the labor market. Globalization, itself abetted by the rise of digital technology, has added a billion people to the world economy in the past generation.[25] The impact of their arrival has been felt most keenly in manufacturing. In 2000, manufacturing jobs in the U.S. employed 17 million human beings.[26] Today the number is about 12 million.[27] Many of the jobs that moved abroad are in low- or middle-skill positions that are themselves now under threat from automation. At the same time, the digital economy has not directly replaced the lost jobs in the United States. A company like Facebook counts more than a billion active daily users but employs only 14,495 people.[28] Twitter has a mere 3,860 employees—internationally.[29]

There is some suspicion that global trade may be running out of steam and that its relentless expansion is slowing down. Technological developments such as three-dimensional printing and the Internet of Things may reduce supply-chain costs and lead to a renaissance of domestic manufacturing. If this is the case, then any new manufacturing jobs will not be open to low-skilled laborers but will require high levels of education and technological savvy. To universities, this is an enormous opportunity to serve learners throughout their working lives. To learners, it is a compelling reason to find ways to distinguish themselves from machines. And unless our machines evolve to surpass our capacity for creativity and mental flexibility, our most powerful skill is our unique ability to be creative. Therefore, we should educate ourselves to do it well—especially considering the unpredictable nature of work.

When the Great Recession struck in 2007, more Americans stayed out of work for longer than in previous downturns. Not until mid-2014 did the employment rate return to its prerecession levels, and although the numbers have rebounded, these are not the same jobs. Many of the new positions are either high-wage professional jobs that require extensive training or low-wage, part-time work.[30] Some Americans have responded to this polarized landscape of economic opportunity by turning to new technological tools, joining the "gig economy" as freelancers through websites like Amazon Mechanical Turk and apps such as Uber. They earn money by performing short online tasks or driving a car for a few hours a day.

In theory, "gig economy" jobs give people the autonomy and the freedom to earn in proportion to the time they invest. Some studies suggest that many people prefer the idea of freelance work for its flexibility and work-life balance. For example, one found that 86 percent of survey respondents at least "somewhat agree" that they would like to work independently.[31]

Analysts, however, say that most people who make money from these new labor platforms are earning only supplemental income. They do not make enough to support themselves.[32] Furthermore, participants in the gig economy lack the security of salaried positions, and many of the jobs they do, such as driving cars and performing routine tasks, are precisely those at highest risk from automation in the near future—witness self-driving cars.[33]

As such, just as previous generations of people turned to education to help them master the economic exigencies brought on by technology, it is once more a very good time to go back to college.

AN EDUCATIONAL FIX

In the past, education has been the surest antidote to displacement by automation. An unemployed weaver could learn to operate machinery. A displaced machinist could learn engineering or management. This upward path was always available because even as lower-skill jobs vanished, economies grew more complex, and so did the work that powered them. Ever-higher skill sets commanded ever-richer salaries. This dynamic is still borne out in the age of intelligent machines. The difference is that with the explosive growth of technology, the educational incline is getting steeper, and universities have a duty to meet this growing demand for learning.

A generation ago, a person could spend four years of her life earning a bachelor's degree and confidently expect entry into a lifetime of steady middle-class employment. This is no longer the case. The pressures of automation and globalization and the increasing complexities of available work have led to stagnating pay among college graduates.[34] College-educated professionals still command an enormous wage premium over high school graduates, but the depressed value of routine labor means that their pay is no longer rising as fast as before. When college-educated workers take jobs for which they are overqualified, they put even more downward pressure on the less educated, driving down wages further at the bottom.

The workers who command the most value on the market today are those with advanced degrees—particularly ones that equip them to work alongside intelligent machines in highly technical areas such as bioinformatics or cybersecurity. Recognizing this, people have responded by signing up for classes. From 2000 to 2010, postbaccalaureate enrollment in the United States soared by 36 percent, while between 2014 and 2025, it is projected to surge an additional 21 percent to 3.5 million students.[35]

But the fact remains that machines will keep getting better at performing skilled work. Consequently, many people are recognizing that education needs to transform into a lifelong pursuit that enables them to upskill and retrain continuously as they try to stay a step ahead of the job-eating robots. Coding boot camps, for example, have seen an enormous rise in popularity among college graduates, with enrollment skyrocketing by 138 percent in 2014 alone.[36] Even so, the day when neural networks simply program themselves may not be much further off than the day when self-driving cars steamroll Uber drivers out of the gig economy.[37]

Just as Uber drivers face a future of disruption, so will many other workers. For the time being, though, lifelong learning remains the surest answer to a long, resilient, and remunerative career. But its form and content may not be quite what the past has taught us to expect. As a matter of fact, a truly useful lifelong education may take a different form from anything that has preceded it.

THINKING DIFFERENTLY

Just as our ancestors could not compete with a steam engine in pulling a load of coal along a railway track, we cannot compete with thinking machines for their sheer brainpower and computational heft. In 1996, Garry Kasparov could not outthink Deep Blue, the chess-playing IBM supercomputer, and since then machines have had twenty years of exponential growth in processing power. As such, the most useful education for today's age will not teach people just how to calculate chess moves or pull metaphorical coal. It will teach people how to do what machines cannot. This means educating people to think in ways that cannot be imitated by networks of machines.

Until now, keeping ahead of technology meant escalating levels of education. The ability to read a handbook once qualified you to operate a mechanical loom; a high school diploma was all the schooling you needed for a lifetime on the factory floor. A college degree was once enough to put you behind a manager's desk, while a master of business administration or law degree opened the doors to the boardroom and the corner office. Look at the LinkedIn profiles of successful tech workers today, and you often will find that they have a master's degree in information technology or project management. But because machines are becoming exponentially smarter, we will need more than simply greater amounts of education to keep pace.

Nor will we simply need education in the content that currently is in vogue among employers. One of higher education's primary purposes has always been to impart content, but intelligent machines are upending the utility of simply knowing things. Information is now instant, ubiquitous, and free. As a result, we need an education that teaches people to learn throughout their lives, bolstering their talents to do what machines cannot.

Which raises a question: what are human beings singularly good at doing? Compared to other animals, we have enormous brains and a knack for digital manipulation that makes us deft with sharpened stones or computer keyboards. But unlike economic eras of the past, we no longer are comparing ourselves to other animals. Robots and advanced machines will soon surpass our most obvious evolutionary strengths, dwarfing us in cognition, precision, and power. But human beings also have evolved as supremely social animals. To survive, our offspring required the social bonds of family and tribe and the imprint of learned knowledge—in other words, of education. This mental flexibility—the ability to learn to speak Mandarin, to catch antelope, or to ride a bicycle—is perhaps our species' greatest survival tactic. At an early age, we can learn almost anything and adapt to any cultural circumstance.

Another result of our sociability is what the historian Yuval Noah Harari, channeling Lewis Carroll, calls "the ability to believe six impossible things before breakfast."[38] We can invent, communicate, and buy into social fictions and abstract concepts (such as money) that unite us and allow us to work together in vast numbers, far exceeding the social capacities of other animals. These fictions can be myths, religions, or ideologies; they can be ideas like human rights, market economics, or national identities. The unique power of these fictions is that they enable us to cooperate on scales vastly larger—to the point of abstraction—than those of our genetic groups or physical communities.[39]

In other words, we have evolved to imagine. We have evolved to be creative. Other animals apply intelligence to solving problems: crows fashion tools to pluck bugs out of wood, and sea otters wield rocks to crack clamshells. But only human beings are able to create imaginary stories, invent works of art, and even construct carefully reasoned theories explaining perceived reality. Only human beings can look at the moon and see a goddess or step on it say we are taking a leap for all mankind. Creativity combined with mental flexibility has made us unique—and the most successful species on the planet.[40] They will continue to be how we distinguish ourselves as individual actors in the economy. Whatever the field or profession, the most important work that human beings perform will be its creative work. That is why our education should teach us how to do it well.

As it has throughout its history, higher education has an important role to play in preparing people for active, engaged lives within society. But as before, it must reflect society's demands. Increasingly, society will demand graduates who possess a heightened power for thinking creatively and flexibly—for thinking differently than machines. Universities already possess an extremely powerful system for teaching this way of thinking. As we have seen, for many decades already, colleges and universities have functioned as loci for creativity. For generations, their research has driven social and economic progress by creating knowledge and translating it into real solutions. It is something higher education institutions do extremely well. Thus, they are ideally positioned to transfer the creative tenets of their research mission with their educational one, using them to help students develop the mental capacity to create new knowledge.

To master the economic and societal challenges brought on by robots, AI, and advanced machines, higher education must continue to keep abreast of change. We cannot educate students as we did in the early decades of the twenty-first century. But if our goal is to train people of the next generation to apply their inherent human strengths to work in the digital economy, universities will have to update their own skill sets. To train both students and current employees for the jobs of tomorrow, universities will have to adapt.

The exact nature of this adaptation depends in large part on the exact nature of tomorrow's jobs. And as the next chapter shows, to find out what that might be, there is no better source than today's managers and CEOs.

VIEWS FROM THE C-SUITE: WHAT EMPLOYERS WANT, IN THEIR OWN WORDS

The town of Westwood, Massachusetts, rests in a green suburban cocoon at the southern edge of Boston's I-95 ring road. A place of shaded streets, well-manicured lawns, and minivans with bumper stickers touting the town's sports teams, Westwood regularly appears on lists of "best" places to live in the state. It is the sort of town that appears to be a stronghold of U.S. socioeconomic prosperity— a bastion of middle-class values, appearances, and incomes. It is a town for the gainfully employed. And in a newly erected shopping mall adjacent to the roaring interstate and the Amtrak station, it is home to a 135,000-square-foot Target megastore.

Target, with its walls and signage an instantly recognizable shade of bold red, is among the breed of "big box" retailers that have become synonymous with the globalized economy. Under a single roof, these stores sell the entire contents of twenty-first-century American life, from baby clothes to patio furniture to prescription drugs. Most important, they do it cheaply. They are able to fill American closets with inexpensive athletic wear and American living rooms with inexpensive televisions because their products are the endpoint of an intricate mesh of supply chains that span the breadth of the world. The teenager earning $10 per hour as a cashier at Target tucks his phone assembled in Shenzen, China, by laborers earning about $17 per day[1] into pants made by Bangladeshi garment stitchers whose minimum wage is as little as $68 per month.[2] As consumers in a consumer economy, all of us are inextricably bound in these complex networks of financing, design, production, transportation, marketing, and consumption.

This inexpensive consumption has its price. Although you might not guess it from looking at the overflowing shopping carts and the new Japanese SUVs in the parking lot outside Target, more than one-third of Americans believe that globalization has been a mostly harmful phenomenon.[3] From calls to build walls along the U.S. border to the "Brexit" vote that will remove the United Kingdom from the European Union, this backlash against the global economy is one of the defining aspects of recent world politics, fueling populist rage against "global elites" and rattling our contemporary political and economic orders.

In the United States, much of the anger is a result of the hollowing out of the middle class. Whereas middle-class households once comprised a majority of people in the United States, members of the middle class are now outnumbered by members of the combined lower- and upper-income households. This has bifurcated the country, pushing its economic and social climate to what the Pew Center calls "a tipping point."[4] Although in previous generations, the economy possessed an expansive middle, over recent years, it has become increasingly hourglass-shaped. At the upper end, in 2015, the top 10 percent of earners in the United States took home more than half of all income in 2015. Meanwhile, the top 1 percent—those earning an average of $1.4 million—took home 22 percent.[5] And although wealth at the top has grown, more people have slipped to the bottom. Since 1971, the share of adults living in lower-income households rose from 25 percent to 29 percent.[6] Economic inequality presently has become one of the defining qualities of American life.

But globalization is not the only culprit. The Great Recession, the decline in the power of labor unions, and automation all have taken their toll on the middle class. According to the Pew Center, middle-income households today are those that, after adjusting for size, earn an income two-thirds to double that of the U.S. median. In 2014, this meant that a three-person household would have to draw in between $42,000 to $126,000 annually.[7] However, the manufacturing economy that once provided many of the jobs that fall within that income range has now gone the way of Pontiacs and Oldsmobiles. In its place, we have seen a massive shift to the service economy, often in the form of jobs with lower wages and fewer benefits than in the past.

For example, on the U.S. Bureau of Labor Statistics' list of the thirty fastest-growing professions, only twelve fall within a salary range that would meet the definition of middle-income. Among the ones that do, there is a healthy representation of technically oriented and hybrid careers, such as computer software and systems software engineer, database administrator, computer systems analyst, and network systems analyst. But most of the expanding professions, including the booming fields of home and personal care aides, earn wages that fall far short of placing their members in the middle-income bracket.[8] And although the gig economy offers people wide opportunities to augment their wages with freelance work, for most families, juggling multiple forms of employment to stay afloat is not an optimal choice.

The difference between a person's ability to obtain a middle-income job versus a lower-income one is largely a matter of his or her qualifications. This makes the middle class increasingly—perhaps inexorably—the province of the college educated, a trend that has been holding steady ever since the dawn of the global economy. Since 1980, the number of U.S. jobs that require more than a high school education has grown by 68 percent. This is more than double the rate of job growth in fields requiring less training or education.[9] Simply put, to obtain and keep a foothold in the middle class—to say nothing of stepping up to the upper echelon of the economy—it seems that Americans will require more and more knowledge.

Yet even this time-tested premise may be coming into question, given the rise of robots and advanced machines. In the past, even as technology put some employees out of work, the economy generated new jobs. Displaced employees could fill these jobs by getting further education and training and acquiring greater amounts of knowledge than they needed for their obsolete positions. Today, however, as intelligent machines move into the workplace, the correlation between knowledge and value in the labor market is shifting. Knowledge economy sectors like finance and law are feeling the impact of machines that perform knowledge work. Some jobs that once offered salaried employment are moving into the gig economy. In the digital age, it seems that even high-paying, prestigious jobs are not safe.

In a world increasingly driven by computers, software, and algorithms, people who have knowledge in these domains remain in strong demand. For example, at a recent job recruitment event put on by Facebook, it was clear who was being sought. As the presenters spoke about optimization, petitioning algorithms, and data fetching, they handed out information cards stating, "Who are we looking for? Bachelors, Masters and PhD students who are studying Computer Science (or related subject)."

"Anyone with strength in data and analytics, you can apply," explained one of the Facebook representatives. "But if you have coding experience, it's a big plus."

"When you join us," added another presenter, "you get involved in every change in the world."

Scenes like this augur well for college students studying high-tech subjects, as well as current employees skilled in such fields. But does it also mean that those of us who do not possess such skills are doomed to an economically inferior future? In a roboticized world, will we, unlike the students recruited at the session that day, be shut out from our chance to change the world?

As it turns out, not necessarily. According to a 2016 survey of employers, the skill cited as most desirable in recent college graduates is the very human quality of "leadership." More than 80 percent of respondents said they looked for evidence of leadership on candidates' résumés, followed by "ability to work in a team" at nearly 79 percent.[10] These are both social skills that people develop through real-world interactions with others. They are also, until someone instills a computer with the commanding presence of Winston Churchill or the coalition-building skills of James Madison, not vulnerable to automation. Written communication and problem solving—skills more commonly attributed to a liberal arts education than a purely technical one—clocked in next at 70 percent. Curiously, technical skills ranked in the middle of the survey, below strong work ethic or initiative.

At the same time, many employer surveys skew to pessimism about today's employees. It is common, for example, to hear companies cite the "skills gap," in which new hires lack the training to cope with the demands of a fast-paced modern work environment.[11] Some economists dismiss this perspective as groundless, citing the stagnant median wage as evidence that the job market is a buyer's one and that employers feel no compulsion to raise wages to attract talent. But that only proves that there is stagnant demand for the commonplace skills of median employees. *Premium* employees possess skills that command premium wages. As a matter of fact, since the 1980s, the compensation for the top-paid 10 percent of people in the workforce has risen sharply in comparison to those at the median.[12] Regardless of compensation, premium jobs—meaning the most fulfilling, creative work in either the private or public sectors—are plums for the most qualified.

There also are openings at the lower bulge of the economic hourglass. At the front of the Target in Westwood, a display board informs shoppers that the store is hiring. Customer service associates, as always, are wanted.

WORKING WITH MACHINES

In late 2016, the White House's National Science and Technology Council's Committee on Technology released a report titled "Preparing for the Future of Artificial Intelligence." In its heavily footnoted fifty-eight pages, the report offers policy recommendations for dealing with machines' imminent capacity to "reach and exceed human performance on more and more tasks."[13] As the report ominously notes, "In a dystopian vision of this process, these *super-intelligent* machines would exceed the ability of humanity to understand or control. If computers could exert control over many critical systems, the result could be havoc, with humans no longer in control of their destiny at best and extinct at worst."[14]

The report observes that the implications of an AI-suffused world are enormous—especially for the people who work at jobs that soon will be outsourced to artificially intelligent machines. Although it predicts that AI ultimately will expand the U.S. economy, it also notes that "because AI has the potential to eliminate or drive down wages of some jobs ... AI-driven automation will increase the wage gap between less-educated and more-educated workers, potentially increasing economic inequality." The report ends with a recommendation for further study of the matter.

Such a study might start by examining how technology is already transforming the workplace, changing the nature of skills even in sectors that traditionally have been insulated from automation. For example, the banking business is now comprised largely not of accounting tables but of complex computer models. According to David Julian, executive vice president at Wells Fargo, one of the largest retail banks in the United States, "We have enormous models that have enormous implications to how we manage our business. We've got millions of loans, and some system has to calculate the interest."[15] At a more sophisticated level, he adds, a computer model may attempt to predict losses in the housing market in ten years' time by tallying and analyzing quantities of data that are unfathomable by the human mind.

Indeed, today's banks no longer need just tellers and accountants. Increasingly, they need engineers and data scientists. They need to construct complex computer models and understand their inner mechanics so they can test them. "We can easily test input, data into a system," said Julian. "We can test the data that we get back. But it's hard to test that black box. And there's so much more reliance on that black box. I've had to hire a lot more folks with background skills who understand how to open up the box to see if it's doing what it's supposed to be doing."

Those background skills are similar to the ones that might appear on the résumé of a Facebook employee. "It's both math and technology majors," Julian said of his recruits. "You have to have folks who can build the models, but more importantly, it's the math majors who can figure out correlation."

This shift to a reliance on computer models has banks scrambling for talent. Julian's risk management team has grown from 550 employees to 950 in three years, and he expects to hire another 100 to 200 per year: "They are a very, very sought after commodity. The geeks are the rock stars in financial institutions right now."

Perhaps the fastest-rising rock stars are the geeks designing the software that already has replaced swaths of Wall Street analysts and traders. For instance, a 2016 *New York Times Magazine* article describes Kensho, a software company that analyzes and predicts the performance of investments by digesting enormous sets of data.[15] It can do this far more quickly and reliably than a team of human analysts, making its creators and investors very wealthy. The fortunes of the displaced bankers are another story.

Even so, according to Julian, demand for technologically skilled employees at U.S. banks is outpacing domestic supply. Wells Fargo actively recruits around the world: "Quite frankly, we're not the only company obviously trying to hire them, so we're over in India a lot more, we're down in the Philippines. We're trying to bring people in."

Scott Semel, the CEO of Intralinks, a legal technology company, has seen a similar shift in the legal profession—one that most people might think is "robot-proof." In the past decade or so, law firms have outsourced much of the yeoman's work of the legal profession—research, fact checking, citation cross-referencing, and the like—to countries such as India, which has a strong tradition of professional legal training. More recently, however, this outsourcing has given way to complete automation at the hands of artificial intelligence. For instance, Semel has said, many contract lawyers now use AI to verify the accuracy of research on the revenue of a company involved in an acquisition. The software undertakes the laborious task of double-checking 80 percent of a company's revenue, while human lawyers spot-check the remaining 20 percent. According to Semel, "You can red-flag things using AI that you simply couldn't get done without an army of associates," and AI does a better job: "The lower-level kind of legal work, those guys should be worried."[17]

Prior to AI, clients paid handsomely for teams of associates to conduct discovery by individually reading each letter and e-mail. "You would have an army of young lawyers in diligence popping up red flags," said Semel. Today, search engines have replaced this function: "No one wants to pay for lawyers to read through a million documents. They're going to do a keyword search."

Not all keyword searches are created equal, however. Joe Basile, a partner at Foley Hoag, a leading U.S. law firm, has observed that the choice of search terms can yield drastically different results and that the person typing the words still requires linguistic skills, analytical ability, and a deep understanding of the underlying law. According to Basile, "A lawyer who didn't make the right connections would do a partial job. They would come up with a bunch of cases but potentially overlook something relevant." Lawyers today still have to understand legal principles, draw analogies, make analytical connections, and give advice. To put it another way, they have to exercise sound judgment.

As a result, Semel and Basile believe that law firms are staking more of their fortunes on the provision of higher-level advice from senior lawyers. "Firms that cultivate that ability to make experienced judgments and provide high-level advice are highly valued," said Semel. Lawyers are increasingly freed from the annoyance of routine tasks so they can perform complex, even creative, cognitive work. To established lawyers, this is exciting and welcome. Entry-level employees, on the other hand, lose the chance to practice law at a more basic stratum and accumulate experience and institutional knowledge. For recent graduates, it may seem like a window of opportunity slammed shut.

Perhaps the most far-reaching impact of all these changes is downward pressure on the wages of lawyers as a profession. "Law firms could still have plenty of associates," said Semel. "The question is, how much are they going to make? Maybe they don't make $160,000 or whatever to go work in a law firm. It happened in medicine." As in all matters of economics, it is a question of supply and demand. Routine legal services have become cheaper because automation has increased the supply of labor. On the other hand, the high-level legal counsel proffered by seasoned professionals still commands a robust hourly rate. Thus, until machines learn to give wise counsel, the higher-level legal services will still be in demand—and those who provide it will still find lucrative work.

Like law, other professions that once seemed far from the technological sphere are also reacting to its encroachment. For example, today's media companies, advertising firms, and marketers have enlisted the power of big data and advanced software to maximize page views and reach more human eyeballs. "The media industry is run by robots," said Grant Theron, executive vice president of global production and partnerships at advertising and marketing giant Young & Rubicam. "It runs on computers and algorithms and targeting."[18]

William Manfredi, executive vice president of global talent management at Young & Rubicam, agrees that marketing today has essentially been transformed into a process of data analytics: "It's about how you interpret the data to understand people's behavior. What's the insight? What's the 'creative' around that, and then what are the channels?"[19]

Increasingly, these channels are automated. "The expression of the creative idea is slowly being pulled into an automated space," said Theron. When you tap into your device, you are not fully in control of your experience. The technology knows tremendous amounts of information about you, including your consumer choices and where you live, allowing it to customize what you see instantly. "The gap between creativity and actually getting something on screen is slowly being swallowed up by technology," he said.

"It's more of a science than ever before," agreed Manfredi. "And the people who can see that process from end to end: they are going to be driving the business in the future."

WORKING WITH SYSTEMS

In contrast to law and advertising, technical sectors like heavy manufacturing might seem susceptible to automation from top to bottom. When Pete McCabe was vice president of global services organization for GE Transportation, he oversaw the services side of a branch of the venerable multinational that constructs, deploys, and manages heavy transportation machinery—massive railroad engines and the like. Much of McCabe's purview was the stuff of complicated software. For instance, his organization manages the flow of traffic on 800-mile-long stretches of single-track railway, determining when to shunt trains to the side to optimize delivery times. According to McCabe, "We have some very, very sophisticated algorithms to drive up to a 10 percent change in velocity, and improvement in on-time schedules and deliveries. The difference of one mile per hour for a small railroad is worth $200 million. For a big railroad it's $400 to $500 million."[20]

Over the past decades, industrial titans such as GE increasingly have moved their business strategies away from hardware and toward software. Instead of staking their fortunes on the sale of giant trains or jet engines or industrial turbines, they now also generate revenue from monitoring them, diagnosing them, and optimizing their performance. McCabe cited the example of a hospital CT scanner. In the past, he said, "you had a sophisticated supply chain, and then you had sophisticated people who could troubleshoot. When a piece of equipment broke, you sent somebody out. They analyzed it, they replaced a part, and got the machine back up and running."

That was before the emergence of diagnostic software. Today, GE operates "what looks like a NASA control center," McCabe said. "You have a room in Waukesha, Wisconsin" in which technicians monitor the machine constantly, watching its "harmonic signature" against its historical norm to spot any deviations. If a deviation occurs, algorithms check it against other variables to determine if the deviation is abnormal. They then can proactively adjust the software settings to attempt to fix the problem before the machine breaks. "In the best case, we eliminate all unscheduled downtime," said McCabe. In other words, GE makes an increasing share of money nowadays not from selling products or replacing them but from monitoring oceans of data to keep its products running.

The shift in emphasis from hardware to software has cut costs, improved efficiencies, and upended the skill sets needed by employees. In the past, McCabe said, "99 percent of your workforce was mechanically trained—a trade school type education in electronics, or mechanics, or HVAC. People who could go out, and diagnose and repair these pieces." In contrast, today, he is hiring engineers, data scientists, and software programmers: "I've got a two-hundred-person software team. In 2000, or even in the 1990s, I had none."

Nonetheless, McCabe added, the skills needed by these two hundred software specialists do not begin and end with a fluency in C++. Domain-specific training—even in high-demand domains like data science—is not enough. According to McCabe, "The problems that are going to change outcomes fundamentally, whether it be in productivity, or health care, or wherever else, are going to be systems problems. It's the interconnection points of all the discrete problems inside a function. We've optimized the hell out of them, and there's always more room, but the exponential value is delivered when you start connecting the dots."

For instance, GE recently set out to address the problem of "wind blow-over derailments." On open, blustery stretches of the Great Plains, ferocious gusts of wind occasionally hit boxcars at a 90-degree angle, pummeling them clear off the track. This threatens public and environmental safety, ruins schedules, damages the company's reputation, and wrecks the bottom line. But McCabe knew that GE's power business had created a model for predicting which trees might fall over in a storm, helping utility companies position their repair trucks in advance. Thus, despite the lack of anemometers on the desolate flats of Nebraska, his group decided to apply the concept to predicting wind blow-over derailments.

"We said, 'In ninety days, we're going to build a prototype and test it and validate the hypothesis,'" said McCabe. Software engineers set to writing applications and interfacing with core systems. Data scientists started looking at wind velocities. "It's the kind of crazy thing that nobody conceived of yesterday, and no single function could solve, or no single kind of profession. The software guys can't do this by themselves. It really is cross-functional."

To oversee systemwide projects like this, McCabe was always looking for the Holy Grail of employees—what he calls "the quarterback." "I can find engineers, I can find software guys, and I can find good data scientists," he said. It is harder to find someone who can draw all the threads together to oversee the team of specialists: "Knowing how they plug, knowing where to push. I'd give my left pinky for ten more of those guys."

Holistic, systems thinking is a quality needed in any complex operation. Take the example of building an airplane engine. According to Andrea Cox of GE Aviation's engineering quality and compliance division, an engine contains between five thousand to ten thousand individual parts, of which an engineer can "own" anywhere "between five and a hundred part numbers."[21] In the past, it was sufficient for a materials engineer to understand the material characteristics of the pieces, a manufacturing engineer to understand how the parts are made, and a mechanical engineer to understand how they operate. Only the design engineer needed to understand how an individual part operates in its own design, how it fits into the design of the module, how the module fits into the engine, and how the engine fits into the airplane. Likewise, the vice president of engineering would have to understand how all the engines delivered on customer expectations, how they fit into the company's strategy, and how they set up the next generation of sales.

As the digital age opens the floodgates to vast new oceans of data, however, it now is possible to accumulate ever-greater amounts of information about systems such as airplane engines. Using the data that airplane engines collect about themselves, engineers can now predict the future operation of a specific part much more accurately, factoring in the details of its situation—for instance, if it is being operated in a hot, sandy climate or in a cold, wet one. "Instead of saying the engine has a life of three thousand hours," said Cox, "we can say, with carrier A, you can probably run it for four thousand hours, but carrier B, you need to take it out at twenty-five hundred hours." Instead of being reactive, engineers are now expected to be predictive. They are expected to harness the available data to get better results from the holistic product.

"The expectation," said Cox, "is to look at that part as a whole— how it's going to end up and where and how the customer is going to feel. How can I use that data to make my customer's world better with the hardware?" This sort of holistic thought is valued by all sorts of companies. Indeed, it is what all companies look for in managers.

WORKING WITH IDEAS

More and more, managerial abilities such as cross-functionality are now a requirement for entry-level positions. Some companies, such as Google, even make it a cornerstone of their hiring process. "When you interview for Google, you don't interview for a job," said Steve Vinter, engineering director and site lead at Google's offices in Cambridge, Massachusetts.[22] Vinter said that Google likes to hire generalists. The interview process gauges candidates' responses to broad challenges, rather than quizzing them on any specific area of knowledge: "How do you think? How do you analyze problems? How do you develop algorithms? How do you measure the performance of those algorithms?" Because verbal answers to questions like these reveal only so much, Google's job application process tests candidates through collaborative problem-solving activities. One of the upshots of this hiring system is that it measures candidates' sense of curiosity, their instinct for innovation, and their knack for working well with others. Consequently, the interview process mimics the actual work a candidate would perform if hired.

To get a foot in the door, though, candidates first need to demonstrate advanced technical knowledge and judgment—what Vinter called "fluency and fundamentals." He described fluency as a person's ability to solve a problem for which she previously was trained. For instance, said Vinter, "you can describe the algorithm you're using to solve a problem. You can write code for it that's correct, straight, simple, and clear. You can describe what it does and test it."

Fundamentals, on the other hand, are the understanding of the many data structures that can be applied to a problem. As Vinter described it: "If there are twenty ways to solve a problem and only two or three that are very good, you have to know how to focus on those and creatively think about how to apply them." Analytical reasoning is the basis of both fluency and fundamentals.

Technical expertise and good reasoning skills are not enough, however. Google's company culture also demands a particular temperament. Employees will not succeed unless they are both highly collaborative and socially accountable to each other. For instance, one of the defining features of the Google workday is the "scrum."

"Scrums are basically daily check-ins in which everybody stands around and talks about what they're going to do during the day," explained Vinter. "Sometimes it'll be fifteen people. Some of them are doing things that appear to be unrelated, but you discover their relation by virtue of telling everyone what you're going to work on. You have a chance to say, 'Oh, I know something that's relevant to what you're doing.'"

This structured serendipity is reinforced by employees having to report back to the group about their daily progress: "It creates this team sense of ownership," said Vinter. In other words, the system taps into the fortuitous side of peer pressure but also of conceptualizing, synthesizing, and communicating ideas.

"Demo days" are another feature of Google's operational culture. "Everyone demonstrates what they've done during the week or something that they're thinking about," said Vinter. It is like a show-and-tell for extremely accomplished technologists: "This creates involvement in what other people are doing." As such, Google's system taps into the fortuitous side of employees' pride and inquisitiveness but also their natural tendencies to evaluate the ideas of others and creatively build on them. These very human qualities are a crucial component of Google's astounding success in bridging the gap between the digital and living worlds.

Other technical and professional industries have a similar appetite for the ability to absorb ideas, evaluate them, and apply them productively. Darren Donovan, managing principal at the consulting giant KPMG, has talked about the need for employees to demonstrate "deep listening skills."[23] Likewise, Sjoerd Gehring, global vice president of talent acquisition at Johnson & Johnson, has looked for a honed sense of reasoning in his employees: "Number one is developing a thorough understanding of the demographics, behaviors, and buying values of the user base, target audience, or end user. It is really hard for us to become more digital, more focused on user interface and user experience, and to tell our story if our own employees don't have a sense of what we mean when we talk about those energies. Can you find a consumer parallel to your own life and apply it to the user base we're targeting? What's behind spending so much time designing things in a certain way?"[24]

These are complex questions requiring intellectual discipline and nuanced thought—and the professional workplace of tomorrow is only getting more complex. Soon enough, professionals will function in tandem with intelligent machines. Whatever the industry—finance, law, manufacturing, media, or any other—it will require cognitive capacities that equip it for tasks we might not even be able to imagine yet. These capacities are mindsets rather than bodies of knowledge—mental architecture rather than mental furniture. Going forward, people will still need to know specific bodies of knowledge to be effective in the workplace, but that alone will not be enough when intelligent machines are doing much of the heavy lifting of information. To succeed, tomorrow's employees will have to demonstrate a higher order of thought.

As we have heard from employers, automation is revolutionizing many industries and driving up the value of technological skills in obvious sectors like tech but also in finance and advertising. It seems as certain as the law of natural selection: if people are gifted coders or are versed in the more desirable flavors of math or engineering, they will find the gates of the labor market wide open and welcoming—or they will for the moment. Software is eating the world,[25] so we need software developers. But it is less clear what we will need when software finishes its meal and settles down to digest. What happens when robots learn to program themselves?

Perhaps an instructive example can be found in sectors, such as law, that are responding to automation with an increased call for sophisticated critical thinking. As we have seen, highly developed critical thinking is essential to work in places as varied as accounting, pharmaceuticals, and leading tech firms like Google. Even the most technical of industries, heavy manufacturing, is reliant on hiring good systems thinkers—on people who can comprehend and act on a broad perspective.

A honed capacity for critical and systems thinking is tremendously valuable in today's workplace. It is essential for the workplace of tomorrow. Every team will still need strong players as well as quarterbacks.

CRITICAL AND SYSTEMS THINKING

The definition of *critical thinking* is somewhat fluid, but for the purposes of this book, we can say that it involves analyzing ideas in a skillful way and then applying them in a useful one. To do this well, a person needs to be able to observe, reflect, synthesize, and imagine concepts and information and to communicate the results of the process. In short, critical thinking is the desired end product of much of what we do in education.

Machines are getting better at many of the elements that fall under the umbrella of critical thinking, including observation and communication. But they have not grasped all of them. Thus, when a lawyer mulls a thorny contract dispute and figures out how to position a client for a victory and when a marketer crafts the content of a website that engages a target audience and keeps eyes on the screen, they are using cognitive capacities that are exclusively human. Critical thinking will therefore remain a cornerstone of human work in the digital age.

Similarly, systems thinking involves seeing across areas that machines might be able to comprehend individually but that they cannot analyze in an integrated way, as a whole. Why certain hashtags on Twitter are trending, why the global commodities markets are rising and falling, why the Antarctic ice shelf is melting: all of these are examples of complex systems in action.

Given their intricacy, conceptualizing systems may seem like a task for which digital minds are better suited than human ones, and we do, indeed, rely on computers to understand complex networks. But computers cannot decide what to do with that information. For example, a computer can model climate change, but it takes human beings to devise and enact policies to stem it. Likewise, in the case of wind blow-over derailment, computers can help a team of engineers predict when it is likely to occur, but they cannot marshal the different talents needed for the project, give them direction, interpret the wider ramifications of the findings, and decide how to implement change. Indeed, a computer would not have had the idea for the project in the first place. Only humans exhibit that sort of creativity.

Because critical thinking and systems thinking are crucial for the human employees of the future, it is imperative that we instill them through the education of the present. Universities will have to develop methods to nurture these cognitive capacities in students if they hope to maintain their age-old social compact, equip graduates for fulfilling, productive lives, and generate new knowledge. To compete with intelligent, advanced machines, we will need to think intelligently about advancing higher education.

A LEARNING MODEL FOR THE FUTURE

The season of discontent is not winter. According to the popular professional networking site LinkedIn, it is October, the month in which worldwide job applications spike. In homage to this annual burst of restlessness, LinkedIn delves through its copious data on hiring and recruiting activity every autumn to compile a list of skills that have proved most in demand by employers for that year.[1] According to its 2016 version of "The Top Skills That Can Get You Hired Today and Tomorrow," the most desirable skill in the global job market is "Cloud and Distributed Computing." Next up is "Statistical Analysis and Data Mining," with "Web Architecture and Development Framework" rounding out the top three. The list continues in a similar vein: "Middleware and Integration Software," "User Interface Design," "Network and Information Security." According to LinkedIn's list, every single one of the ten most desirable skills on the planet is technological.

Even as technology-related jobs continue their ascendance and as technology is integrated ever more thoroughly into every industry, the fact remains that technology will not provide jobs for everyone. Even if technological work were available to everyone who is qualified, the advance of robots, advanced machines, and artificial intelligence suggests that ultimately, technology itself will become the best candidate for many jobs. A great many economists, jour-nalists, and thinkers argue that deep learning in machines and the exponential growth of big data and processing power are rendering the human mind an economic relic. In *Rise of the Robots: Technology and the Threat of a Jobless Future*, Silicon Valley entrepreneur and author Martin Ford paints a convincing picture of the tsunami of automation that is about to wash away white-collar jobs. "Exponen-tial progress is now pushing us toward the endgame," he writes. If technology can replace human beings on the job, it will. Pre-venting business owners from adopting labor-saving technology "would require modifying the basic incentives built into the market economy."[2]

Ford and the others are entirely correct that machines will destroy jobs. Historically, this has been the case, going back to the days of the Luddite weavers. Nonetheless, technology also has given rise to new industries, bringing new forms of employment. Once again, this is the case, as can be seen with the invention of jobs that did not exist a generation ago, before the rise of the Internet era, such as search-engine optimization specialists or user-experience designers. But as journalist Ryan Avent observes in his 2016 book, *The Wealth of Humans: Work, Power, and Status in the Twenty-first Century*, "New technologies will create new, good work, which might often benefit the less skilled. But it will not be scal-able mass employment. And it will not solve the problem of labor abundance."[3] Despite the invention of unforeseen jobs, the conflu-ence of "automation, globalization, and the rising productivity of a highly skilled few" will continue to suppress the value of human labor across the globe.[4]

Nonetheless, when these analysts and futurists argue that the current technological revolution is different and that the value of human labor will be irretrievably lost, they are overlooking two salient facts. First, much of the world remains *terra incognita*. There is more to find, in the heavens and on earth, than we can dream of in our blinkered present. We have a universe of scientific secrets to uncover and shoreless oceans of knowledge yet to cross. We have an infinite canvas to paint and endless music to play. From curing disease to restoring the environment to writing the next great novel, there is everything left to do. And for most of us, this involves finding fairly compensated, satisfying employment. Thus, even as machines take over routine labor, freeing us from repetitive or mundane tasks, human beings have a great deal left to occupy them. The only question is whether we possess the tools to accomplish it.

This relates to the second point that many of the current analysts overlook—the historic role played by education in elevating the majority of people to the next level of economic development. In the nineteenth century, free, public elementary schools raised the mass of Americans out of illiteracy. When technological progress built steam, public secondary schooling helped the merely literate ascend to the next rung of the ladder, giving them basic skills they needed to work in the new industries. Then, as corporate America rose in the postwar order, public colleges and universities raised mass education levels another notch again, teaching a sizeable portion of the workforce advanced knowledge skills. Now, once more, technology is raising the educational bar.

If the work of tomorrow demands more from us, we must demand more from our education—particularly at the college level.

Consequently, an education for the digital age needs to focus not just on technology and understanding what technology can do but also on what it *cannot* do—at least for now and perhaps never. In other words, a robot-proof education nurtures our unique capacities as human beings. And the most elevated of all human capacities is the one that may be the most elusive and difficult to define and therefore is trickiest to teach. This is humanity's unique talent for creativity.

THINKING CREATIVELY

Few would contest that Wolfgang Amadeus Mozart was one of the most creative people who has ever lived. The man who wrote Symphony no. 40 in G Minor and *Don Giovanni* is the quintessential example of a creative genius, a person whose talent transcended all boundaries of genre, context, and culture. In 1756, the year of his birth, the human population on earth numbered approximately 795,000,000.[5] Today, the world's population is almost ten times that amount, at 7,000,000,000. Mathematically, we can induce that there are about ten creative geniuses alive today who were born with an extreme but not unique genius for musical composition identical to Mozart's. Chances are, many of them are currently living in China or India.

In addition to having more and more human competition, budding composers and songwriters now have to contend with musical algorithms. For instance, Flow Machines, an artificial intelligence program from Sony's computer science lab in Paris, recently released original jazz and pop compositions. At the time of this writing, it is working on its debut album.[6] The recording industry appears to be getting more crowded.

Education cannot teach everyone to be Mozart. However, in a global, automated economy in which more and more exceptionally talented people compete on the same plane, a useful education will assist those of us without singular abilities in achieving singular outcomes. Education should therefore cultivate our creativity. But to teach creativity, we first have to understand what it actually is.

Cognitive psychology has given us a rich and varied body of scholarship on creativity. In the 1960s, Paul Torrance developed a sequence of tests that attempted to quantify a person's creativity. For example, they might ask a child to draw details around a neutral shape, incorporating it into a picture that tells a story. These tests are still widely used, especially in schools for assessing children for the purposes of gifted education but also in corporate contexts.[7] Yet even the most ardent proponents of the tests agree that creativity is a phenomenon of such astounding complexity that it is extremely hard to tease out its mental constituents.

Nonetheless, one of the most useful concepts on which the tests are built is J. P. Guilford's articulation of convergent and divergent thinking.[8] When a person employs convergent thinking, she focuses on finding the single, "correct" answer to a problem or task. Answering questions on a multiple-choice test is an example of convergent thinking at work. When we use convergent thinking, we weigh data and alternatives to achieve the one best, black-and-white result. This is exactly the type of mental activity at which advanced computers and machines are now becoming adept.

Divergent thinking, on the other hand, is the creative generation of multiple responses in a free flow of ideas. Examples of this include brainstorming and free writing—the outpour of ideas on

49

the page without regard to structure or grammar. Divergent thinking is often associated with playfulness, curiosity, and willingness to take risks.

Divergent and convergent thinking share many of the same elements. For example, both require the ability to assess and elaborate. However, divergent thinking requires creativity—a sensitivity to the changing nuances of a problem, a facility to reframe it as circumstances demand, and ultimately, an ability to generate a result or resolution that contains things that were not there when one started.[9]

Generally speaking, the educational system in the United States—both K–12 and college—focuses primarily on training students to master convergent thinking. When we ask students to determine which of two trains traveling to Chicago from different points at different velocities will arrive first, we are flexing their convergent thinking muscles. Yet this is precisely the sort of thinking that increasingly is a robot's specialty. Even academic challenges that mix convergent and divergent thinking may be automatable. For example, by studying vast numbers of writing samples, algorithmic programs can assign accurate grades to essays written by human students.[10] And with machine learning on the rise, it probably will be no great leap for them to write A+ essays themselves sometime soon.

Divergent thinking, however, is another matter. Whereas machines can grade essays on the causes of the Napoleonic Wars—and may soon be writing them—they cannot produce *War and Peace* starting from a blank screen. To accomplish this type of work, we still need human brains. But our schools often do not do an especially good job of cultivating them.

The most popular TED (Technology, Entertainment, Design) Talk of all time is Sir Ken Robinson's "Do Schools Kill Creativity?" recorded in 2006.[11] In it, he famously argues that creativity, which he defines as "the process of having original ideas that have value," is as important to today's children as literacy. However, by stigmatizing failure and wrong answers in school, we train children to stifle it. "We don't grow into creativity," says Robinson. "We grow out of it, or rather we get educated out of it."

Because the U.S. education system was designed largely to meet the needs of the industrial economy of the nineteenth and twentieth centuries, it tends to emphasize the skills most valuable in a world made up of factories, bureaucracies, and ledger ink. Often, though not exclusively, it pushes students to learn mathematics, language, science, and other "hard skills," with less emphasis on the liberal arts, creative disciplines (such as music and art), and metacognitive skills typically referred to as "soft skills." As Robinson observes, our educational system tends to reify academic performance above all else, such that the ideal product of the educational process, if all the tests are aced and all the coursework assimilated, is a university professor. However, as he correctly notes, this lopsided view of intelligence ignores the richness of human capacity. "Our education has mined our minds like we've strip-mined the earth, for a particular commodity," says Robinson, insisting that there are greater treasures to be uncovered than mere academic performance.

The view that mastery of facts and knowledge is what makes a person "smart" or "prepared" *is* a lopsided view of human intelligence—and never more so than in the present moment, when robots, advanced machines, and AI are increasingly able to master facts and knowledge as effectively as the "smartest" of us. And yet colleges and universities tend to reinforce this lopsided view. Colleges and universities, in their DNA, are mechanisms for the transmission of knowledge about specific subjects. As far back as the invention of universities in the Middle Ages, they organized themselves around the separation of disciplines, with faculties specializing in the content of theology, law, or medicine.

Today, most colleges' curricula and pedagogy still place inordinate weight on the transfer of information into students' minds. Development of students' higher-order mental capacities, like critical thinking or elegant communication, is certainly one of the objectives of a college education, but all too often it is secondary to the ingestion of content. More often than not, college courses are not designed to nurture metacognitive skills explicitly and systematically.

The problem is that even if we aim to teach these skills, we do not necessarily do it well, much less nurture students' creativity. In their 2011 study, *Academically Adrift: Limited Learning on College Campuses*, professors Richard Arum and Josipa Roksa found that "at least" 45 percent of the undergraduates they surveyed showed "exceedingly small or empirically nonexistent" gains in critical thinking, complex reasoning, and written communication during their first two years in college. After four years, 36 percent of their sample still showed no improvement at all: "They might graduate, but they are failing to develop the higher-order cognitive skills that it is widely assumed college students should master."[12]

Other data reinforce this grim prognosis. Under its Program for the International Assessment of Adult Competencies, the Organisation for Economic Co-operation and Development (OECD) conducts surveys of life skills in adults from thirty-three different countries, testing them on literacy, numeracy, and problem solving in a technology-rich environment. Americans have not performed with particular aplomb. For instance, the most recent survey found that 30 percent of Americans aged thirty-four or younger with bachelor's degrees failed to score above two out of a five-level numeracy assessment. For its assessment of problem solving, this dismal figure rose to 34 percent.[13]

You may ask, "What does it matter if students can't think, if machines will increasingly do the thinking for us?" But rather than raise the white flag on humanity and wipe the educational slate clean, we need to reconsider what we teach. If we rebalance our approach—by helping students to acquire the content they need to understand their chosen domain of study, as well as the broader cognitive capacities they need in a highly automated professional workplace—future generations will not be abandoned to the economic dust heap. However, we need a new model of learning that enables learners to understand the highly technological world around them and that simultaneously allows them to transcend it by nurturing the mental and intellectual qualities that are unique to humans—namely, their capacity for creativity and mental flexibility. We can call this model *humanics*.[14]

Much as engineering and philosophy are both disciplines that involve the study of a body of knowledge and the development of a way of thinking, humanics is a discipline that teaches mastery of content as well as the development of particular skills. It helps people understand the components of the technological world while giving them the ability to utilize it, manipulate it, and ultimately transcend it. Humanics is a discipline tailored for our era that is grounded in the mastery of content knowledge I call the *new literacies* and in the development of "robot-proof" ways of thinking I refer to as the *cognitive capacities*.

THE NEW LITERACIES[15]

The word *literacy*, meaning in its plainest sense the ability to read and write, stems from the Latin *littera*, a letter of the alphabet. The term usually encapsulates numeracy, which makes sense considering that human writing originally began as a means of recording mathematical data. Both letters and numbers allow us to represent verbal expression through symbols, so we can preserve them and transmit them to other people. This ability to telegraph ideas into the minds of other people simply by showing them a few abstract marks is one of humanity's most powerful tools. And as anyone who has ever lost themselves in a good book— or a financial spreadsheet—well knows, it also is an early form of virtual reality. Literacy delivers information and ignites the imagination.

Before the Internet, writing was the way that human beings enshrined facts, encoding all the accepted truths about a civilization's history, property rights, genealogy, law, and culture. Mastery of the written word made a person wholly human. It was the basis of a person's ability to live a full life within society. It was power. That is why it sometimes was restricted to privileged groups, with its secrets forcibly withheld from marginalized ones. Frederick Douglass wrote that literacy is the path from slavery to freedom—conversely, a deficit of literacy is a slide into powerlessness.

Literacy gives us the power to network with the ideas and information produced by other people at any distance of time or space. Written language allows us to communicate ideas, mathematics allows us to communicate about quantities and dimensions, and scientific literacy allows us to communicate about the natural world. In a digital milieu, human beings require more complex literacies that enable us to do more than simply transmit concepts between human minds. Humanics' three new literacies—technological, data, and human—enable us to network with both other people and machines. Even more so, they empower us to use the digital world to its fullest potential.

Technological Literacy

The first of these new literacies is technological literacy—knowledge of mathematics, coding, and basic engineering principles. Today's "digital natives" have grown up immersed in digital technologies and possess the technical aptitude to utilize the powers of their devices fully. But although they know which apps to use or which websites to visit, they do not necessarily understand the workings behind the touchscreen. People need technological literacy if they are to understand machines' mechanics and uses. In much the same way as factory workers a hundred years ago needed to understand the basic structures of engines, we need to understand the elemental principles behind our devices. This empowers us to deploy software and hardware to their fullest utility, maximizing our powers to achieve and create.

Because coding is the lingua franca of the digital world, everyone should be conversant in it. But before people can learn the syntax of computers, they need to understand the conceptual elements. It is rapidly becoming the established wisdom among early educators that coding has a place in the classroom next to the ABCs and 123s. Toymakers are increasingly rolling out new products that inculcate the concept of coding in young minds. Within the next generation, it seems inevitable that coding will stake out a place in the required high school curriculum—and perhaps in the college curriculum as well. Indeed, one of the most noted developments in the higher education marketplace in recent years is the rapid growth coding boot camps, such as those operated by General Assembly and Dev Bootcamp. According to a recent market report, coding boot camps are expected to graduate nearly eighteen thousand students in 2016, growing by a vigorous 74 percent over the previous year.[16]

"Coding is as important now as math was ten, twenty, thirty years ago," according to Sjoerd Gehring, the Johnson & Johnson executive. "Of course, the two are very, very connected."[17] Indeed, many efforts like Scratch are underway to democratize coding, wresting its mysteries from a cabal of mathematically gifted programmers. Tools like Ready and Hopscotch are designed for even the mathematically bereft to jump into creating original software. Hailing the rise of this sort of "pop computing," David S. Bennahum, cofounder and CEO of programming tool Ready, writes, "It's only a matter of time before the process of making software itself is transformed, from one that requires a mastery of syntax—the precise stringing of sentences needed to command a computer—to the mastery of logic."[18] Until democratic programming comes to pass, it is essential to gain a grounding in specific computer languages and the basics of computer science.

Data Literacy

One of the consequences of technology's encroachment throughout our lives is the concurrent explosion of data. In 2010, *The Economist* reported that Dutch cows were getting "smart" sensors implanted in their ears to track their health and activities—a sort of conflation of Big Data and Big Brother to create Big Cow. We now hear about the Internet of Things bringing connected, "smart" physical objects to every part of our homes, our clothing, and our environment. By 2020, it is estimated that we will live in a world of fifty billion smart objects, creating a true ecosystem of information.[19] And this is not happening solely in the home. Even pieces of heavy machinery, like earth movers, are no longer simply gears and metal but are equipped with sensors that analyze data and feed it back to the operator to optimize the machine's accuracy and efficiency.[20] We are awash in data, so the next of our new literacies is data literacy, the capacity to understand and utilize Big Data through analysis. By understanding both interpretation and context, data literacy enables us to find meaning in the overwhelming flood of information pouring from our devices.

There is little use in accumulating massive amounts of data unless we can arrange it into usable information and thence into understanding. Data analysis allows us to do this by sifting through these giant sets of data to find the correlations in them that yield useful findings. Indeed, this is the basis of Google's business and that of countless other digital companies today. Based on the correlations we discover, we are able to understand the real meaning of the information and then extrapolate accurate predictions from it. Data analysis can foresee everything from the spread of a virus across a continent to an individual's dating preferences. It is an extraordinarily powerful tool that is getting more powerful every time someone clicks on the Internet. However, to be fully functional actors in our digital world, people need to understand how to use this tool—and to grasp its limitations.

As Michael Patrick Lynch observes in his book *The Internet of Us: Knowing More and Understanding Less in the Age of Big Data,* the information we derive from making correlations with big data often can be misleading unless we understand its context.[21] He cites the example of a famous video map of cultural history that was created by using a data set of births and deaths of "notable" people over the past two thousand years. The shape and findings of the map were entirely dependent on the creators' assumptions of their data's parameters, including what constitutes a "notable" person. In other words, the answers we get are only as meaningful as the questions we frame. And for that, we need an understanding not just of correlations but of how and why the facts are so.[22] Instead of seeing strands of information, we need to view the interconnected tapestry of relationships in a system. Thus, it is not enough to see that "culture" spread to different geographical hotspots in a particular pattern. We would need to study the social, economic, and political contexts. The purpose of data literacy, then, is to give us the tools to read the digital record and also to understand when we ought to look elsewhere.

Human Literacy

The last of the three new literacies is the most important and perhaps requires the least explication—human literacy. Even in the robot age—or perhaps, especially in the robot age—what matters is other people. Human literacy equips us for the social milieu, giving us the power to communicate, engage with others, and tap into our human capacity for grace and beauty. It encompasses the humanities traditionally found in a liberal arts education but also includes elements of the arts, especially design, which is integral to much of digital communication.

Neoclassical poet Alexander Pope wrote that the proper study of mankind is man. He was not entirely correct—the rest of the natural world holds its wonders—but intellectually, morally, and spiritually, the humanities are among the most fertile grounds on which to nurture a complete human being. They form the foundation of a life well-lived and the furnishings of a civilized mind. That is reason enough to study them. But they also happen to be starkly practical. Professionals need a strong grasp of human literacy because despite our digital landscape, we live and interact with humans. Even in a fully networked space, the most powerful networks are personal relationships. Workplaces are, as we saw in the previous chapter with Google's "scrums," more collaborative than ever. In academia, for example, the days of the solo researcher are at an end. As fields become more interdisciplinary and complex and work becomes more hybrid, more and more discovery is undertaken by teams—and the teams are getting bigger.[23] In every workplace context, we have to know how to play well with others, so skills like brainstorming, negotiating, and making collective decisions are increasingly important. Effective relationship work, not just knowledge work, is the key to a winning team.[24]

In this regard, understanding the importance of diversity is essential to human literacy. If students are to be lifelong learners, they must engage with a diversity of perspectives, including ones that challenge their presuppositions. Only through the full and respectful inclusion of people of different backgrounds, identities, and creeds can we learn, cooperate, and create to our full potential. Divided communities are weaker than unified ones, and although the notion of a global community has suffered from a backlash recently, the fact remains that technology, economics, and mass culture bind us closer than at any time in history. No one can afford to pretend that the world, in all its variety, can be locked outside our doors, much less our devices. Instead of dividing us, human diversity is an astonishing source of beauty and strength, proving the limitless nature of humanity. By immersing students in diversity and celebrating its lessons, we enrich their minds, broaden their thinking, and build their valuable human literacy.

In addition to getting along well with other people, we have to know how to communicate and motivate them. We may glean factual information on human behavior through data literacy, showing us the *what* and the *how*, but the humanities teach us the *why*. As the boundaries between technology and the humanities dissolve, even the engineer needs to consider human interfaces, and even the programmer must learn to be a storyteller.

The extension of technology into every aspect of life has very human ramifications that we have to address through politics, economics, law, philosophy, and especially ethics—subjects that must evolve with the growth of AI. Almost ironically, a fundamental aspect of human literacy involves the ethical quandaries raised by intelligent machines. The old trolley problem—do you swerve a moving vehicle into a crowd of bystanders, or do you doom its occupants?—is now a very material question for the makers of autonomous vehicles. The military faces thickets of ethical and legal conundrums as autonomous weapons become a technological reality: What ethical principles should govern the design and development of AI? How do we align these new machines with our values—and which values do we favor? If they cause harm, who is morally culpable? Only human beings can unleash the free agency of machines in situations that will result in human death. To take that step or not, we will need philosophers as well as lawyers.

Another ethical dilemma raised by intelligent machines is the old specter of inequality. In part, this is a continuation of the ongoing polarization of wealth. But in an economy in which more and more labor is performed by robots, more and more of the value of labor adheres to the robots' owners—to capital. Employees lose out even more, so some form of redistribution becomes necessary. Human literacy therefore must steer our social policies, striving to bend the arc of history toward social justice. But another, less familiar facet of inequality is just as pressing. Artificial implants and enhancements will soon be a reality for people who can afford them. By augmenting their human abilities with cybernetic upgrades, some people will boost their lifespans, senses, physiques, and perhaps even intelligence. Furthermore, families that can afford to purchase genetic modifications might snip away bothersome genes (like nearsightedness or a predisposition to weight gain) but also customize sons and daughters to their most exacting specifications. In the future, the rich might truly be different from you and me. The question for ethicists is whether such powers are to be restricted, freely granted, or hoarded by the wealthy. Human literacy will help us make the right choices.

THE COGNITIVE CAPACITIES

The new literacies (technological, data, and human) are the foundation of humanics—its core curriculum, in a sense—but they are not sufficient to educate people to master a highly technologized world. To do that, students also need a higher order of four cognitive capacities that will serve them in the digital economy. As we encountered in the previous chapter, these capacities include critical thinking and systems thinking—metaskills that everyone needs to analyze and apply ideas and to understand and command complex systems. Two other cognitive capacities are necessary to help make learners robot-proof. The first is entrepreneurship—the act of creating value in original ways. The second is cultural agility—a capacity that enables students to operate deftly in a global milieu and to appreciate the varying understandings and values that people from different cultures bring to an issue or situation.

Critical Thinking

Critical thinking is about analyzing ideas skillfully and then applying them fruitfully. Machines are certainly improving in elements of this capacity. Their powers to observe, analyze, and communicate are growing stronger with every upgrade, but they lack the ability to synthesize and imagine. Although a machine might be much better than a human being at using data inputs to tackle a specific problem—winning at chess, organizing a global supply chain, finding you a compatible date for a Saturday evening—they are not as impressive at more unquantifiable thinking.

The act of critical thought involves numerous layers and striations. Some of these are quantifiable forms of thinking, like understanding and applying facts to a question. Others are inchoate, even intuitive, such as envisioning how people's motivations, emotions, and histories influence them. True critical thinking requires all these layers for a full understanding of the context in any situation.

If a problem can be reduced to a train of yes and no questions, no matter how complex, then a machine can resolve it. But many real-world problems defy such reduction. For example, imagine that a social media app sees that its user base is flat. By answering yes and no questions such as, "Are users aged eighteen to thirty-four in Korea decreasing in number?" or "Does this particular ad generate clicks during peak hours?," the computer might be able to crunch all the relevant data about pricing, distribution, marketing, and strategy. It even might be able to identify the exact color of design most likely to catch a user's eye and calculate the precise keywords that would resonate on social media with the target customer base. This is the data analysis side of critical thinking, and computers are getting better at this every day.

Despite this analytical rigor, the computer would not be able to guess at other vital factors that could affect the success of the company's plan just as much as more quantifiable data does. For example, it would not be able to gauge customers' reservoirs of brand loyalty. It would not be able to measure their visceral response to the company's advertising or take into account the personal and cultural associations customers have with particular images. Without this layer of contextual analysis, a machine might go ahead and greenlight a marketing plan that looks good in terms of data and metrics but that, for lack of considering other important contextual factors, will be a flop. In contrast, humans are alone in their ability to assess both sides of the critical thinking coin—data analysis and context—and say, "This plan will or won't work."

Very often, the difference between successful and botched critical thinking boils down to questioning assumptions—choosing to ask if an accepted input is, indeed, correct. For example, on several space shuttle missions, the National Aeronautics and Space Administration observed pieces of foam breaking off the external tanks during launches. It happened again during the doomed *Columbia* launch, with foam striking the left wing. But NASA scientists made the decision that the foam breakage was no cause for concern, unwittingly condemning the shuttle to disintegration on reentry.[25] They operated on their existing assumptions instead of taking the critical leap to wonder, "What if this time things are different?"

That same fault in critical thinking led to the catastrophic 2005 levee failures in New Orleans. For years before Hurricane Katrina struck, studies and investigations had warned of the potential for disaster, with authorities even running exercises in preparation for a category 3 storm over the previous year. But when it actually happened, the city's flood protection system failed, triggering the greatest engineering disaster in U.S. history. The cause was not lack of information but a presumption that the designs were sound. It was a failure of critical imagination among those who held responsibility.

Systems Thinking

Machines are adept at understanding the elements of complex systems and the ways their variables cascade into one another, but they are less skilled at knowing how to apply this information to different contexts. So, for example, a machine can model the impact of climate change on a coastal area, assessing water temperature, pollution, currents, weather patterns, and a host of interweaving factors. By assessing all the data, a machine could yield conclusions about how to improve nearby architecture and combat erosion. But that same machine would not imagine how to deploy the data in different fields like economics, law, or health sciences. It would not, for example, make the decision to use the information in a study on human migration, apply it to operations in the fisheries industry, or write environmental legislation. Locked in the silos of its programming, it would not imagine the value of breaking out of domain-specific thinking.

Computers could be programmed to think across a variety of silos, enabling them to engage in systems thinking of a sort, but the big creative leaps that occur when humans engage in it are as yet unreachable by machines. For example, Dutch architect Koen Olthuis fanned such a systems thinking spark into a creative flame when he began to think about the intersection of climate change, urban planning, and architecture. Understanding that rising sea levels threaten cities around the entire world, he invented floating buildings—ranging from cargo containers buoyed by recycled plastic bottles in Bangladeshi slums to luxury artificial islands in the Maldives—that can withstand the long-term effects of higher seas and blurring coastlines.[26] Olthuis tackled the challenge of climate change through the foundational lens of an architect but also through the simultaneous frameworks of an environmentalist, a materials scientist, an engineer, and a design pioneer to come up with a solution no computer can yet muster. He thought about urban sustainability as a forest of interconnected subjects, not a single disciplinary tree.

Systems thinkers possess the ability to tackle the problems that challenge us most. To address the tragic contaminated drinking water crisis in Flint, Michigan, systems thinkers would not treat it solely as a public health issue. They also would address it from the perspectives of civic infrastructure, taxation, leadership, and justice. Biologists combatting the Zika virus would view it as a medical emergency but also model its progress using network science and think about public outreach campaigns through a marketing lens. In the same vein, economists examining the Great Recession would consider the roles played not only by subprime lending but also by mortgage-backed securities and credit default swaps, failures in government regulation, and macroeconomic conditions.

Systems thinking is a critical cognitive capacity for anyone in a position of leadership but also for anyone attempting to discover new knowledge, launch a business, or create something original. It sees the details and the entire tableau, exercising our mental strength to weigh complexity while also testing our grasp on multiple strands of thought. Educational groups like the Waters Foundation are working to bring such systems thinking exercises to K–12 classrooms, embedding it in language, math, and social studies.[27] Colleges, too, can explicitly draw out the systems thinking lessons in their courses.

Entrepreneurship

As machines invade the labor market, the third cognitive capacity, entrepreneurship, will be increasingly valuable as a means for people to distinguish themselves in the digital workplace. To put it plainly, as machines fill old jobs, we will need to invent new ones. Indeed, because of the technologizing of the workforce, we probably will feel more pressure as a society to do so. One estimate from the World Economic Forum claims that 65 percent of children entering primary school today will eventually work in jobs that do not yet exist.[28] These jobs will be invented by entrepreneurs who strive to push the boundaries of discovery and invention, as well as to generate wealth. From this point of view, technology is not a threat but a source of opportunity. It does not destroy jobs; it generates potential new ones. The distinction is a matter of entrepreneurship.

This is one of the most compelling reasons that entrepreneurship should be a baseline capacity for all college learners. This capacity functions in two dimensions. The first dimension is the traditional startup model. As machines fill our existing roles in the labor market, we need to think of new roles in which we can expand that market by launching new ventures and new industries. The second dimension functions within the context of established institutions and businesses. Employees with an innovative mindset will invent new ways to bring value to their companies and new fields that no technology can yet master. In this way, entrepreneurial energies are reformative. For example, it was through an entrepreneurial mindset that GE's management pushed the firm to reinvent itself from a twentieth-century manufacturing company into one that now focuses primarily on twenty-first-century technology and services. This shift fundamentally reinvented the company but did not start a business from scratch. These two sides of the entrepreneurial coin—startup and evolution—harness the same drive for creation but latch it onto different entities.

As a matter of fact, entrepreneurship is a capacity that can be applied to any business context, including ones that have no intention of generating profits. Entrepreneurs who bring original ideas to bear on social inequities can use the tools of the marketplace to ease poverty, boost development, and advance social justice. Teaching entrepreneurship—especially social entrepreneurship—should thus be a matter of national consequence and a priority for universities.

Yet despite popular appearances, such as the cultural prominence of celebrity entrepreneurs, entrepreneurship rates in the United States are down. Since 1994, the number of Americans employed in newly launched businesses has dropped by more than one million.[29] As a society, we have to turn this trend around, and our success depends on whether our cultural and social climate is conducive to supporting innovation, change, and experimentation. The whole world may not be Silicon Valley, but the whole world can be inspired by it.

Desh Deshpande is the founder of Sycamore Networks and the Deshpande Foundation, a nonprofit aimed at accelerating entrepreneurship for social and economic impact. In 2010, President Barack Obama appointed him as the cochair of the National Advisory Council on Innovation and Entrepreneurship, and he has given a great deal of thought to how the entrepreneurial mindset functions.

According to Deshpande, "There are three types of people in the world. There are some people who are oblivious to everything, some people who see a problem and complain, and some people who see a problem and get excited to fix it. The difference between a vibrant community and an impoverished community is the mix of those people."[30] The need for a preponderance of entrepreneurial people is even more pronounced as advanced machines proliferate in the workplace. People will have to think of creative ways to work.

Universities, with their critical masses of active minds, are ideal entrepreneurial ecosystems. In addition to developing academic programs that teach entrepreneurship, higher education can support initiatives that empower students to experiment with business ideas. "The entrepreneur's journey has to be experienced," said Deshpande. "It's a bit like having kids. You can talk all about what it's like, but unless you do it, you'll never know what that experience is like."

A large part of that experience is failure. Machines, unlike people, are designed to always succeed at their tasks. If there is a systems failure of some sort, the consequences are usually unpleasant. Yet when human beings fail, the consequences can sometimes be providential. For example, Alexander Fleming famously discovered penicillin after accidentally contaminating a petri dish with mold. Entrepreneurship requires an acceptance and a sideways view of failure. The saying "Fail fast, fail often" has now become a mantra in Silicon Valley. FailCon, a global conference for tech entrepreneurs to relate their tales of loss, is held everywhere from Toulouse to Tel Aviv.[31] The term "failing upward" has entered the tech lexicon, and a corpus of business literature extols the lessons of noble failure. Yet in the business world, those lessons come at the heavy human cost of lost investments and curdled dreams. The beauty of entrepreneurship education in universities is that students can learn those lessons before they incur the real pain of ruined credit, unpaid bills, and broken relationships.

Learning from failure is also an integral part of the scientific method lesson that students grasp through original research projects. Research is a form of intellectual entrepreneurship in that it is the creation of valuable knowledge. It teaches critical thinking, systems thinking, and creativity, pushing students to contextualize their ideas within the framework of existing knowledge and then to imagine new avenues of discovery. For example, a student studying environmental science might work on researching changes to coastal ocean temperatures. In addition to creating a method for gathering data and drawing conclusions from her findings, she would have to consider the ramifications of her work within the larger system of coastal sustainability. She would exercise her ability to think in detail, in broad strokes, and in new dimensions.

Cultural Agility

Experience is also an essential component in the final of our cognitive capacities—cultural agility. This is, according to my colleague Paula Caligiuri, "the mega-competency that enables professionals to perform successfully in cross-cultural situations."[32] In the past, cultural agility might have seemed mostly the provenance of jet-setting business travelers and diplomats, but globalization has vaulted this skill into a mainstream imperative. Furthermore, no matter how accurate iTranslate becomes, true cultural agility is beyond the reach of machines. It requires empathy, discretion, and a very human nuance—protocol droids like *Star Wars*' C-3PO will remain a figment of Hollywood for a long time to come.

Cultural agility involves more than just knowing how to behave in a video conference or at a foreign restaurant. It requires a deep enough immersion in a culture so that we can fit seamlessly into multicultural teams or get results from people who have dramatically different lives from our own. It also is applicable closer to home. Organizations have different corporate cultures and expect different behaviors from their employees, so it is a useful skill when switching between jobs. Consider, for example, the potential for disconnect between professionals moving from a casual California tech startup to a hierarchical Wall Street firm.

So, too, the flow of digital commerce is resetting workplace norms. Although international trade in goods stalled after the 2008 downturn, digital flows—which comprise a snowballing portion of GDP—have increased dramatically. A 2016 McKinsey report found that the amount of cross-border bandwidth used globally has grown by forty-five times since 2005.[33] A full 12 percent of global goods trade is now e-commerce. Increasing amounts of the traffic are performed between machines. One estimate suggests that, by 2019, 40 percent of global connections between devices will be performed by machines with other machines.[34] Even so, humans are fully immersed in this global movement. The McKinsey study found that 361 million people e-shop across international borders, 44 million work online in other nations, and fully 914 million cross borders in their social networks.[35]

Increased global commerce means increased complexity in business dealings due to cultural differences. And increased complexity means increased chances of misunderstanding. Even simple assumptions may be fraught. For instance, a few years ago there was a popular trend for challenging people to donate money to ALS research or dump a bucket of ice water over themselves. In the United States, this was seen as a harmless meme that helped raise donations for an important cause. In India, it was regarded as a shocking waste of clean water, so participants replaced it with the idea of donating a bag of rice to someone in need.[36]

We all need to understand, empathize, and collaborate with people from starkly different backgrounds. As of 2014, foreign-born inhabitants of the United States numbered a historic high of 42.4 million, or 13.3 percent of the total population. If you count the U.S.-born children of immigrants, that number rises to 81 million, or 26 percent.[37] In our culturally diverse economy, the most successful professionals will be those who can step lightly across divides, showing psychological ease in making decisions in different contexts, integrating or adapting, and succeeding at different roles.[38]

There is not a machine in the world that can advise us on how to respond to a stranger at a crowded bar or how to react to the vocal tone and body language of a new colleague (although we have seen enormous leaps in software that can read facial expressions).[39] Machines might be able to tell us the exact meaning of a businessman's words, but they would not be able to negotiate a deal while accounting for subtexts, unspoken meanings, and cultural assumptions.

Cultural agility is crucial for problem-solving across borders. A marketer selling an automobile would invent completely different publicity campaigns for Dubuque or Dubai. Likewise, a medical administrator in New England could call for the purchase of an expensive magnetic resonance imaging machine for her hospital, but the same machine might prove useless in another context—say, a clinic in Bangladesh that does not have the necessary electricity grid or maintenance support. In such a case, the best answer would be reverse innovation—looking at how to fill a need cheaply or with the resources available in the developing world. In the case of the clinic, it might mean using inexpensive, battery-powered medical devices.

Context is everything—and that context is not easily appreciated by even the most intelligent of machines. The globalized economy has lifted the borders on business but not on the multiplicity of contexts in which human beings live and work. Until advanced machines learn to navigate the infinite variety of human belief and behavior, humans will continue to be the masters of our shared intercultural milieu.

BUT HOW TO TEACH IT?

If the goal of higher education is simply to insert information into a student's brain, a library card or Internet connection would be the only tool we need. But most people are not autodidacts, and most college students do not master the content of their degree programs simply by reading. We employ seminars, essay assignments, exams, and a multiplicity of other tools to teach academic content. Thus, when we rebalance the objective of a college education away from its current overemphasis on content delivery and toward teaching the new literacies and cognitive capacities, we likewise need to expand our pedagogical toolbox. This involves thematic study across disciplines, project-based learning, and real-world connections.

In thematic study, instructors can turn an implicit process of learning into an explicit one. A traditional liberal arts program, for instance, might see professors leading students through rigorous consideration of gender in the Victorian novel while implicitly teaching strong writing and critical thinking skills. In our new model, teachers have to expose the underlying fabric of learning to their students, like turning a sweater inside out. They need to delineate clearly what is being studied, practiced, and acquired, explicitly identifying process and goals in every component of a course. This does not necessarily require subscribing to all the prescriptions of the assessment movement in higher education, with its emphasis on student testing and institutional metrics. But it does mean that instructors should be explicit about their goals.

As such, their syllabi ought to describe the four cognitive capacities developed through each step of study and discussion. Each lecture would include a clear learning outcome along these lines. Exercises, exams, and simulations would be designed with similar goals. So, for example, a program teaching sustainability would map out how each assignment builds students' systems thinking skills while exposing students to relevant concepts from environmental science to data analysis, urban planning, physics, and health sciences. A business professor would explain how a case study imparts lessons on entrepreneurship. An assignment in a class on international contract law would denote how it develops cultural agility.

Furthermore, students would not consider their subjects in silos but would undertake hands-on projects that give them the opportunity to synthesize knowledge across different fields. They might, for example, apply the lessons of a business course and a computer science one to launch a tech startup out of their dorm room.

Another tool available to instructors stands just outside the classroom door. Because the world itself is the most effective teacher, educators also should be more mindful about connecting their teaching of the four cognitive capacities to applications outside the academic cloister. Students who practice systems thinking need to understand how their mental skills can serve them in real life. In this instance, perhaps, their instructor could help them see how the ability to assess a complex problem through multiple interdisciplinary lenses will help them manage a project team in a nonprofit fighting climate change. The key is to enable students to understand how, exactly, their acquisition of the new literacies and development of the cognitive capacities will serve them in their life goals—not simply as scores on a transcript.

These tools—explicit learning across disciplines, project-based learning, and real-world connections—are imperative in teaching the robot-proof model of higher education. But if the lessons are frozen in the classroom environment, they lose immediacy. It is one thing to hear a story told; it is another to act as the protagonist. Human beings, unlike robots, acquire their deepest programming through their senses as they explore the fullness of their environments and of their lives. Unlike machines, our greatest teacher is experience.

4

THE EXPERIENTIAL DIFFERENCE

In 2011, IBM's supercomputer, Watson, became the first machine ever to compete on the game show *Jeopardy!* Its opponents were the two greatest human champions in the game's history—Ken Jennings, who held the record for the longest winning streak, and Brad Rutter, who was the show's biggest all-time money winner. But although Watson soundly defeated his biological competitors, it did not play an entirely flawless game.

In one clue, host Alex Trebek said, "It was the anatomical oddity of U.S. gymnast George Eyser, who won a gold medal on the parallel bars in 1904." Jennings incorrectly answered that Eyser was missing an arm. Watson then chimed in, "What is a leg?" Although it guessed the body part correctly, it failed to note that the limb was, in fact, missing. David Ferrucci, who led the Watson project, observed that the computer had probably tripped up on the word "oddity." It would not have known, he said, that a missing leg is necessarily odd.[1]

Its biggest mistake, however, occurred in the final round of day 2. Under the "U.S. Cities" category, the clue read: "Its largest airport was named for a World War II hero; its second-largest, for a World War II battle." Both of the human beings correctly answered, "What is Chicago?" But Watson had been programmed to focus on the precise wording of the clue, giving less weight to the category title. Because "U.S. city" never appeared in the phrasing, it incorrectly guessed, "What is Toronto???"

Despite this hiccup, Watson handily routed the humans, and in the aftermath Jennings published an essay in *Slate* describing his defeat. "Just as factory jobs were eliminated in the 20th century by new assembly-line robots," he wrote, "Brad and I were the first knowledge-industry workers put out of work by the new generation of 'thinking' machines. 'Quiz show contestant' may be the first job made redundant by Watson, but I'm sure it won't be the last."[2]

Jennings was certainly correct, but a more intriguing observation came during a moment of consolation. One of the IBM engineers told Jennings that his streak of game show victories had helped inspire the Watson project. "We looked at your games over and over, your style of play," the engineer said. "There's a lot of you in Watson."

Machine learning, as the engineer's comments suggest, is similar to human learning in some ways and very different in others. Computers such as Watson improve by casting their nets across inputs, trawling through them for patterns. The computer picks the "correct" answer based on the broadest consensus of its data sets. In many ways, it is a display of informational crowdsourcing. In Watson's case, some of those inputs included Jennings' play style and strategies.

Intelligent machines get better through exposure to wider and wider pools of information. At a certain point, this learning technique is not unavailable to us because human brains cannot possibly absorb data at such a scale. But human brains have their advantages, too.

The human *Jeopardy!* players erred through lack of knowledge, sluggish reflexes, nerves, or perhaps simply a hiccup of the synapses. Watson's mistakes, however, were due to context—or lack thereof. In the case of Eyser's leg, a human being, having observed other mostly bipedal human beings on a near-constant basis for her entire conscious life, instantly would have known that it is atypical to be missing a limb. Likewise, in making an error most schoolchildren would have avoided in guessing Toronto, Watson did not show a lapse in its elementary geographic knowledge. Rather, it failed to think outside the parameters of the exact question. It failed to recognize context.

Computers have a problem with interpreting contexts because they do not live in the chaos of the human world, even though they do straddle the digital and physical environments. As computers equipped with sensors become more and more adept, they increasingly are able to operate in and learn from the realities of the city street, the workplace, and the home. But they cannot experience human life—and because they cannot know the world through human experience, they cannot fully account for or fully appreciate human contexts.

In the previous chapter, we saw that an education in humanics—based in the three new literacies and designed to develop the four cognitive capacities—is an evolution beyond traditional higher education teaching and learning. However, it still is not enough to make learners robot-proof for their entire careers. To catalyze their mastery of humanics, people need experiential learning in the contexts of different live—rather than classroom-based or digital—environments. As machine learning advances, computers will continue to grow more sophisticated cognitive capacities such as critical thinking, systems thinking, and even cultural agility. But they will lack the very human lens from which we view life, learning to interpret contexts to assess, act, and make sound decisions. Human beings possess this lens because we learn from experience.

Intelligent machines learn when their artificial neurons form stronger or weaker connections by gradually altering through trial and error. Human beings learn in much the same way, with our synaptic connections getting stronger or weaker. The key difference is that instead of pure numerical data inputs, we strengthen or weaken our mental connections through experience. In chapter 3, we saw that human learning works best when it follows a progression of steps: learners should understand that they are actively building a cognitive capacity, and they need to understand how to apply that capacity. Most important, they then should actually apply it, actively using it in real-world situations while assessing the results and reflecting on the consequences. The goal of experiential learning is to remove the boundaries between the classroom and real life, creating a constant, multidimensional learning ecosystem. This steeps learners in randomness, in the serendipity and weirdness of life that diverts the brain down unmapped channels. It gives them the chance to improvise in contexts they never have encountered before, interacting, inventing, and thinking on their feet. When human learners are immersed in the incalculable variety of experience, they escape the strictures of predetermined input—which computers cannot do. They break free of their programming, and they upgrade their minds.

WHAT IS EXPERIENTIAL LEARNING?

Experiential learning is a model unlike any traditional format in that it integrates classroom and real-world experiences. It flings open the gates of the campus and makes the entire world a potential classroom, library, or laboratory. Typically, students engage in experiential learning through internships, co-ops, work-study jobs, global experiences, and original research opportunities. However, its principles extend to any situation that does not involve the passive absorption of information. In other words, if you do something outside of an academic environment and you learn as you do it, you are engaged in experiential learning.

As we all know, practice makes perfect. But the power of experiential learning is that it places practice in novel contexts. Experience means contact with and observation of facts and events. To make an obvious point, learning occurs when we act and think. Experiential learning's potency, however, is greatly amplified when it purposefully integrates the self, humanics, and the real world. Learners follow an ongoing process of checking, testing, and refining their knowledge, but they do not do it in a random way. Genuine integration is paramount. There must be a two-way street between the application of classroom learning in the context of life and the application of real-world knowledge in the context of the classroom. If this is done purposefully—if it is done *mindfully*—learners peel back the layers of assumption or habit that cloud their insights into themselves. In clear light, they see their abilities, their present skills and knowledge, their predilections, and their room for growth. Consequently, as they better understand the world, they better understand their own minds.

In many ways, experiential learning is life's most sophisticated engine for personalized education. Because learners' experiences are molded by the unique contexts of their lives, they are learning in ways unavailable to any machine. By constantly adapting to shifting reality, learners make unexpected connections. They find inspiration in places they never would have thought to Google, stretching creative muscles and mental flexibility. Through this process, they become more robot-proof.

But how exactly do experience and learning connect? The value of experience in education has been debated for millennia. Some thinkers hold that context is a critical aspect of learning. For example, rationalists contend that reason alone is sufficient means to arrive at the truth about immutable realities, but empiricists argue that knowledge of reality must stem from observations incurred by the senses—from the context of experience. The great American educator and philosopher John Dewey thought that the validity of ideas should be tested against human experience, so he argued for a form of schooling based on such experiences. A student is not a blank slate, he wrote. Rather, he builds on the accretion of previous experiences and knowledge: "What he has learned in the way of knowledge and skill in one situation becomes an instrument of understanding and dealing effectively with the situations which follow. The process goes on as long as life and learning continue."[3]

In contrast to Dewey, traditionalist educators rejected experience as the basis of knowledge, demanding a return to "the logic of ultimate first principles expressed in the logic of Aristotle and St. Thomas." Dewey thought this choice was "so out of touch with all the conditions of modern life that I believe it is folly to seek salvation in this direction." Instead, he called for lived experience to form the grounds of an education, with the "systematic utilization of scientific method as the pattern and ideal of intelligent exploration and exploitation of the potentialities inherent in experience."[4] In other words, he sought to take education out of the context of the closed mind and open it to the entirety of the world.

Dewey and his ideas on experiential education inspired many followers. In the 1970s, educational theorists Ronald Fry and David Kolb followed up on his notions by articulating "an integrative framework for understanding the teaching-learning process."[5] They devised a four-stage cycle based around the process of checking, testing, and refining. In the first stage, the learner begins by doing—by having an immediate, concrete experience. This leads to a second stage of observation and reflection. The third stage is thinking—forming abstract concepts and generalizations. The final stage is planning—testing the implications of these concepts in new settings. The cycle repeats, and learning continues.

For example, an investor in a tech company sees her stock price rise. She observes that this happened immediately after the company announced the launch of a new product. Next, it dawns on her that the company plans to launch another product in six months—so this, too, may boost the stock price. Lastly, she determines to see if her investment in a competing company gets a boost immediately after it launches a rival product.

WHY IS EXPERIENTIAL LEARNING EFFECTIVE?

Although practice does bring learners closer to perfection, a violinist, for example, will not become a virtuoso simply by running a bow along the strings of her instrument *ad infinitum.* To be effective, her learning must follow a structured sequence. Learning science tells us that to master any complex subject, learners must first acquire component skills.[6] Second, they need to practice integrating them into a given context. Third, they need to apply what they have learned in different contexts.[7] So, for instance, an aspiring chef first learns basic knife skills, culinary terminology, and the preparation of stocks and sauces in a classroom. Next, she integrates these components by practicing recipes in a culinary school. Finally, she applies this knowledge by working the dinner shift at a busy restaurant.

The result of this sequence—acquisition, integration, application —is expertise. We can think of a student's progress from ignorance to mastery as an advancement through four stages of development within the dimensions of consciousness and competence.[8] In the first stage, students are unconsciously incompetent. They lack the knowledge to realize the extent of what they do not know. In the second stage, as the extent of this begins to dawn on them and they understand they have much to learn, they advance to a state of conscious incompetence. Further advancing, they reach a state of conscious competence in which they can perform well but must do so with deliberation and intent. At the final stage, they achieve the liberating state of unconscious competence, instinctively operating at the highest level in their domain. So carrying on the example of our chef, she enrolls in culinary school without having ever heard of a mirepoix. Next, she begins to fathom how many stocks and sauces she has never tasted, much less attempted to make from scratch. Later, after months of practice, she finally is able to cook an acceptable beef Wellington. Finally, the day comes when she can run a full kitchen while absent-mindedly whipping up a dessert soufflé.

To achieve a high level of mastery, a chef—or any learner—must first follow the aforementioned sequence of acquisition of skills or knowledge, integration, and application. Acquisition and integration can be undertaken in academic settings. So returning to our robot-proof learning model, students acquire content knowledge such as the new literacies in the classroom. They integrate them through controlled assignments such as writing essays, completing projects, or conducting laboratory experiments. But when they apply them to a novel context, they actually achieve mastery. This is where experiential learning comes into play. Experiential learning is effective because it completes the three-part learning sequence, giving learners the opportunity to take the components they have integrated and apply them to complex, living contexts.

Application is the crucial, final step in the sequence, and its operating principle is *transfer*. Transfer occurs when skills or knowledge are learned in one context and the student successfully applies them to another. If the contexts are similar—say, when students take ideas they learned in a class on Elizabethan drama and apply them to one on Restoration poetry—the transfer is *near*. The students take a theory, concept, or body of knowledge and put it to work in a new but largely familiar situation. If the contexts are largely disparate—for example, when critical thinking skills honed in a Restoration poetry seminar are used to create a public relations campaign for a marketing company—the transfer is *far*. The students are encountering an entirely novel situation but are able to step back and understand how, embedded in the context, they can use their knowledge to solve a problem.

Some educators argue that far transfer is the ultimate goal of education. Learning is not much use, after all, if its efficacy sputters out the instant it leaves the classroom. But far transfer is difficult to do. Studies have shown that students rarely exhibit the ability to apply relevant learning to unfamiliar situations.[9] They may find themselves overly dependent on familiar contexts and inflexible to new applications. They also may lack a deep understanding of their domain, knowing the *what* but not the *why*. This blinds them from seeing how their knowledge could be utilized in a different setting.

The combination of theoretical knowledge that crosses contexts with lived experience is the key to overcoming these difficulties, helping students become adept at far transfer.[10] For example, in a study cited by the authors of *How Learning Works*, a seminal book in learning science, researchers asked two groups of students to throw darts at targets placed one foot under water.[11] Both groups got better at hitting the targets with practice. Then teachers instructed one group about the principle of light refraction. When the targets were readjusted to a different depth, the students who had learned the abstract concept were better able to adapt their throwing techniques, scoring much higher than the ignorant group. In other words, academic or theoretical knowledge enabled them to transfer their experiential knowledge to a new context successfully.

Exercises that prompt students to consider problems in different contexts can help build transfer, as can immediate feedback from instructors or supervisors. This is especially effective in situations in which learners face real-life consequences, such as workplace environments. For instance, our chef may have learned in culinary school how to whisk together Hollandaise sauce, but a sous chef might remind her to avoid overheating it while working a busy brunch service. The lesson is cemented under the pressure of living action. She incorporates the feedback, successfully plates the eggs Benedict, and improves her mastery.

WHY DOES EXPERIENTIAL LEARNING MAKE YOU ROBOT-PROOF?

We have seen that when learners put their knowledge into practice in real-life situations, they develop a better understanding of themselves, their strengths and weaknesses, and their drives and possibilities. They also sharpen their cognitive capacities, leading to the robot-proof qualities of creativity and mental flexibility—both aspects of far transfer. By contrast, no computer has yet displayed creativity, entrepreneurialism, or cultural agility. And although machines are continually improving in their ability to map knowledge onto recognizable problems—in other words, improving in their near transfer abilities—they cannot perform far transfer well, at least not in the infinite contexts of real life.

Some enterprising machines in the future might launch their own financial services consultancy, resolve a fraught negotiation over intellectual property between American and Chinese lawyers, or post an original video that garners a hundred million likes. But they have not done these things yet, and they do not appear to able to do so anytime soon. Human beings need to use their robot-proof cognitive capacities for those purposes. In other words, our potential to master far transfer is our competitive advantage over intelligent machines.

Practicing far transfer stretches not only the mind but also the *mindset*. Psychologist Carol Dweck has articulated a concept of mindset that cuts to the core of why experiential learning is power-ful. According to Dweck, people respond to situations with either a "fixed mindset" or a "growth mindset." The fixed mindset is unable to view adverse contexts as anything but impediments. In this mindset, people believe that their qualities and capabilities are set and immovable. Their thinking is rigid: you are either intelligent, outgoing, or good at math, or you're not. This mindset is reinforced in schoolchildren when we punish failure by awarding bad grades. In many traditional classrooms, the swing between punishment and reward teaches kids to value success and affirmation above anything else. They learn to see setbacks not as opportunities but as personal failures. People stuck in this mindset tend to believe that circumstance is a fence that cannot be breached. They think that great artists, athletes, and scholars are born that way or get lucky. It is a constrained point of view that consequently limits people's potential.

A growth mindset, on the other hand, believes you can change the contexts in which you find yourself—even simply the context of your own thinking. Personal qualities are mutable. From this point of view, adversity is not a negative context but an opportunity for learning and improvement. For example, you may not be particularly outgoing, but you can choose to step up and make the introductions at a cocktail party. You may not be good at math right now, but you can keep practicing your calculus problems.

Growth mindsets see that natural talent and circumstance are not anchored contexts but are merely starting points from which to improve. People with a growth mindset possess the conviction that they can, though effort and diligence, change their abilities. They become self-reliant. This is of obvious use in converting setbacks into future successes, although people with this mindset do not see success as merely winning. To them, the real value in any context is the opportunity to learn.

In other words, the growth mindset posits that in general, situations, contexts, and the state of one's own learning are not inherently good or bad. Their value lies in how we think about them. The growth mindset is essential to nurturing the cognitive capacities of critical thinking and systems thinking because both demand that students cast the nets of their minds on wide, and often unexplored, waters. And it is the key to becoming the most robot-proof person of all—the self-directed, lifelong learner.

Dweck observes that this mindset is also the foundation of something else. "In a poll of 143 creativity researchers," she writes, "there was wide agreement about the number one ingredient in creative achievement. And it was exactly the kind of perseverance and resilience produced by the growth mindset."[12] Indeed, we already have glimpsed this in our earlier discussion of divergent versus convergent thinking. The questions that arise, then, are how do you teach students to identify when they are stuck in a fixed mindset, spinning their wheels, and what sort of mental boost will push them back onto the track of the growth mindset?

Classroom instruction can help. Evidence shows that simply pointing out the idea of mindsets to students compels them to take notice of it and make efforts to change. But reading about Dweck's theory in a traditional learning context is not sufficient. To exercise the growth mindset and build a student's creativity, she has to experience using it. She has to learn by doing.

Experience is the catalyst for the suprarational aspects of learning. By experiencing different situations and contexts, we trigger our emotions, challenge our beliefs, and test the fabric of our minds. These almost subconscious elements of thinking spur our minds to grow—and these subconscious elements cannot be mimicked by computer processors. Deep learning in machines works by scanning data for patterns. Experiential learning in humans works through exposure to the full universe of stimuli. The outcomes are as complex and incalculable as the effects of a rainstorm on a woodland or the flow of water through a riverbed.

EXPERIENTIAL LEARNING THROUGH CO-OP

For college students, one of the most direct forms of experiential learning is cooperative education—an educational model in which students alternate their classroom learning with sustained, full-time immersion in the professional workplace and then integrate the two. "Co-op," as it often is called, is an approach with a long history in institutions of higher education. It began as the brainchild of Herman Schneider, an architect, engineer, and educator who first instituted the model soon after joining the faculty of the University of Cincinnati in 1903. A few years later, Northeastern University adopted the model. Over time, Northeastern and other co-op universities have developed a robust institutional architecture to support this way of learning, integrating it into academic departments and curricula. Establishing and maintaining these is no small task.

Northeastern's co-op program is built on a century of deliberate cultivation, advances in learning science, trial and error, serendipity, and elbow grease. When it began in the first decade of the twentieth century, it provided work experience to fledgling automobile mechanics and electrical engineers. Today, it is a full-fledged experiential learning model that includes partnerships with 3,300 employers in more than 130 countries around the world—including the occasional placement in Antarctica.

This global character is fundamental to the program's success. We believe that the best way to educate students to understand the world—and ultimately, to change the world—is to immerse them in it. As such, we offer students opportunities to live, learn, and gain professional experience in as wide a variety of countries, companies, and institutions as they can. By immersing themselves in different cultures, proving themselves in different professional settings, and experiencing different problems, challenges, and understandings of societal issues, our students gain a deeper understanding of the world, the subjects they are studying, and themselves. When they return to the university from their co-op experiences, they apply all of this in their subsequent academic learning.

Co-ops typically last six months, after which students return to the academic nest. Because most Northeastern undergraduates pursue co-ops—indeed, most complete multiple co-ops—the program needs to work for a wide variety of disciplinary majors. Some students seek out co-ops that hew closely to the subjects they are studying, such as the business major who does a co-op at a financial services firm. Others look for co-ops that are at the leading edge of their fields, such as the health sciences major who seeks out a co-op at a medical center that is implementing a "personalized medicine" model of care. Still others seek out co-ops that are far from their field of study, such as the design major who pursues a co-op with NASA. They may do this out of sheer curiosity and love of learning or because they have an interesting idea about how their discipline and the focus of their co-op may interrelate.

The process of researching, interviewing for, and undertaking a co-op helps many students decide what they want to study or pursue professionally. Others decide, after completing a co-op in a certain field, that the area they *thought* they wanted to study or pursue as a career actually is not for them, after all. Still other students are burgeoning entrepreneurs who use co-ops to understand the area in which they want to launch a business—and then do just that after their co-op is complete.

To match students with co-ops, the university maintains a sizable network of coordinators who work with both students and employers, as well as career development coordinators who work with students. These dedicated staff members perform a twofold function. First, they help students seek out co-ops, determine learning outcomes on the job, and reflect productively on the experiences after their completion. Northeastern students must apply and be accepted to a co-op, and not all students are accepted for the opportunities for which they apply. This in itself becomes a learning experience for students to assess and reflect on as they prepare to apply to another co-op.

Second, our coordinators collaborate with employers to ensure that they provide a meaningful, high-quality experience for the student. Likewise, they work to ensure that students will provide solid value for employers by defining the deliverables and outcomes that the company, agency, or business concern seeks from the co-op student.

A typical undergraduate embarks on her first co-op during sophomore year. After having the established some foundational knowledge in her discipline, she will enroll in a preparatory class that establishes practical skills such as writing a résumé and sitting for a job interview, honing in on her career interests and goals, and developing job-search strategies. Class exercises also expose her to principles of mindful learning so that she enters her co-op with an open mindset. The idea is for the student to approach the professional environment of a co-op with a clear view of how her experiences there will connect with her classroom learning, as well as a deliberate idea of what she wishes to achieve.

After the student discovers a good fit, she and the supervisor at her co-op location delineate her exact job responsibilities and learning objectives. By applying their classroom learning to workplace tasks, students in co-ops repeatedly practice far transfer, cementing the cognitive capacities—critical and systems thinking, entrepreneurship, and cultural agility. Furthermore, they experience the tangible consequences of their actions. In a classroom setting, a lack of preparation or a failure to think three steps ahead might result in a flunked quiz or botched experiment. In a workplace, it could undermine the bottom line, to say nothing of adult lives and careers.

At the conclusion of six months, the student returns to campus to share her experiences living, learning, and working in her co-op with her peers, writing essays analyzing her experiences, and bringing her stories to discussion groups. This debriefing is an essential component of the model. It helps students consciously integrate what they have learned in their professional environment, as well as in their broader immersion in the world, into their course of study and their lives on campus.

Ultimately, a co-op is as different from an internship as a narrative poem is from a haiku. For the students, co-ops are deep and sustained learning experiences, charged with purpose. After a co-op, students better understand their academic disciplines, the rhythms and nuances of the professional workplace, and the part of the world in which they have lived and worked. Most important, they better understand themselves.

There also is compelling empirical evidence that co-ops help students develop a wide range of higher-order skills that make them more robot-proof. In 2015, Northeastern conducted a scientifically designed survey of one thousand employers across twenty-five industries, including manufacturing, science and technology, finance, and insurance. Some of the employers had hired recent Northeastern graduates (all of whom had completed co-ops), while others had not.

We then asked the employers for their perceptions of recent graduates' skills. The results were eye-opening—and demonstrate the power of experiential learning. For example, across a spectrum of nine higher-order skills—including critical thinking, analytical reasoning, problem solving, obtaining and processing information, and working with diverse groups—employers rated Northeastern graduates significantly higher than graduates from other colleges and universities. Likewise, across a range of eleven additional attributes—including leadership, initiative, teamwork, flexibility, willingness to learn, and creativity—employers again rated recent graduates of our university well ahead of those who presumably had not had the benefit of co-ops or other experiential learning opportunities. All of these differences were statistically significant.[13]

It is not surprising, then, that when employers are asked whether co-op and experiential learning models should be more prevalent in higher education, they respond resoundingly in the affirmative. For example, in a national study Northeastern conducted of C-suite executives and business leaders, 96 percent said that integrating education programs with practical experience was important. Nearly half reported that learning new skills and industry-relevant competencies is the most important thing new employees should do in their first five years on the job.[14] Consequently, learning models like co-op that begin this process before a person joins the professional workforce full-time are advantageous. The general public in the United States sees widespread benefits to co-op and experiential learning, too. In a separate national study, more than 85 percent of respondents agreed that the co-op model helps students develop more applied skills, develops job candidates who are better prepared for the real world, and better prepares college graduates to find professional employment in today's job market.[15]

CO-OP IN ACTION

To see how co-op and experiential learning translates into the development of robot-proof skills in human terms, consider Catherine Erdelyi, a math and business administration major at Northeastern. She arrived at college planning to explore a career as a math teacher, but after her first experience volunteering in the community, she realized that path was not to her taste, so she switched to a more analytical direction. After her faculty adviser suggested she look at career possibilities as an actuary, she went on a co-op with a company that developed software used by insurance companies to model risk. "Mostly what I did was work with catastrophe bonds," Catherine told me. "Investors would pay into this bond thinking that if a hurricane is not going to happen, they would get interest from it but that if it does happen, they lose a lot of money. So we would figure out at what threshold they would lose some of their money or all of their money, to determine the risk of the bond."[16] In short, she applied her critical thinking skills to a complex, real-life project, transferring the math skills she acquired in the classroom to the faraway context of the workplace.

In subsequent co-ops with companies such as Liberty Mutual Insurance and John Hancock, Catherine employed systems thinking to interpret the results from computer models. "I worked in the catastrophe management space," she explained, describing how she used the models to determine the losses the company would incur if, for example, a hurricane hit Florida. "My job as an analyst was to know how to interpret the data, who to tell, and why it's important. I had to send it up the chain, so if our loss was going to be X amount, to make sure we were prepared for that. I would have to consider, from a strategic point of view, if that meant buying reinsurance or maybe stopping insuring so many homes in Florida."

Catherine's systems thinking meant the difference between profit and loss, success or failure in very vivid terms. No algorithm could have mimicked her mental flexibility in assessing the entire field of questions, events, numbers, and strategies. No computer could have taken such a broad view of the situation and improvised the correct decisions within the live context of a working company. Through these experiences, Catherine not only practiced her cognitive capacities but also learned to be a more flexible thinker and brought enormous value to her employers.

Or consider the example of Mackenzie Jones, a political science and international affairs major at my university. During a co-op with the Organized Crime and Corruption Reporting Project in Bosnia-Herzegovina, she was assigned to monitor the daily news in the Balkans and other parts of Europe and Central Asia, looking for stories that would fit the organization's mission. If she found a story of interest, she might be tasked with writing a brief on it. But first she had to research its veracity—not always a straightforward job when speaking with people from the contexts of different cultures. "With the Internet, you never really know what you can trust," observed Mackenzie. "I was expected to call news organizations and be like, 'Hi, does anyone in your organization speak English, and could you tell me what happened?'"[17]

A machine mind—namely, Google Translate—helped Mackenzie in her work. But it could not help her ascertain whether a story was true or a total fabrication. Thus, although a computer might easily aggregate news stories and translate them, it would show poor judgment in weighing their merit. In this important regard, Mackenzie had a competitive advantage over any piece of software.

"You have to be careful because if you misinterpret something and repeat it and then you spread the lie, we would lose credibility," she said. "Also, it's illegal to defame somebody." Recently, the proliferation of "fake news" on social media has shown the danger of mixing false information with technology unchecked by any gatekeepers. Technology lacks the discernment to filter truth from deceit. And although human judgment can be grossly flawed, humans are also adept at sniffing out the truth by using context, such as people's visible and not-so-visible motivations, to ascertain what are facts and what are lies.

This sense of nuance and relativity is an important part of cultural agility. A receptive, adaptable mindset is also the wellspring of empathy—a commodity much in need in our increasingly global and technologically complex era. For example, Mary Tobin, a finance and political science major, spent five weeks in Switzerland at the United Nations in Geneva, studying disarmament diplomacy. "I got to meet many ambassadors to the U.N. from a lot of different countries," she said. "People working toward creating regulations around the use of mines or mine cleanup in places like Cambodia that have really awful problems with people still getting legs or arms blown off because of mines that have been there for decades."[18]

Her immersion in these difficult questions led Mary to pursue a research project on human rights, international law, and the use of autonomous robotic weapons. "There's no human actually interacting with the decision to pull the trigger on another human, and [there are] not even any verification systems. Studying this was a really enlightening experience." By interacting with people from places directly affected by tragedies that often are encountered only in news feeds, Mary was inspired to delve deeper into subjects of enormous legal complexity and moral nuance. She explored the idea that while some of us fret about technology threatening our livelihoods, others harbor genuine terror about software weighing the scales of life or death.

Empathy is what led Ali Matalon, a business and political science major with a concentration in entrepreneurship and innovation, to apply her classroom lessons to a seemingly intractable problem. On a co-op with a nonprofit in Jamaica, Ali witnessed the desperate social and economic situation of unemployed young people.

Wanting to help train them for employability and connect them with paid work, Ali and her colleagues had to think entrepreneurially about what resources they had at their disposal. Her project recruited sixteen- to twenty-nine-year-olds from communities at high risk from gang culture. The majority of them lacked a high school education. They had few prospects and little opportunity to break the cycle of poverty. What they did possess, however, were smartphones.

"Ninety percent of them have worked with Word, PowerPoint, or even Excel at some point," said Ali. "They have used things like WhatsApp or Facebook or a number of other technology applications. Those skills become very easily transferable."[19]

Applying the entrepreneurship training she received in class, Ali helped launch a microwork center to assist the young Jamaicans in developing their coding and software skills. Then, using business processing outsourcing (BPO), they set up the trainees to accept contract work. "It's data management and manipulation services," she said. "Someone might need a blog updated, so they throw that up on the BPO system to be sent out as a job. Or a consulting group might need an Excel file sorted but doesn't want to run the sorting software on it because shifts in the data could cause a problem. So humans deal with that work."

Through innovative thinking and entrepreneurship, Ali and her colleagues created a novel approach to tackling a complex challenge. And although this sort of digital piecemeal work does not pay much, the increments accumulate enough to make a difference in people's lives. "This is the nose to the grindstone, get it done—not the most exciting work," said Ali. "But it helps to pay the bills." Hiring young Jamaican contractors for low-skill technology work is cheaper and more efficient than trying to automate it. Technology, in this case, is not destroying jobs but is helping to lift people out of poverty while inspiring students like Ali to use their talents for the greater good. Social entrepreneurship co-ops such as this are a powerful reminder of our common humanity and a powerful way to practice the robot-proof cognitive capacities.

LEVERAGING EXPERIENTIAL LEARNING IN THE RESIDENTIAL UNIVERSITY

Northeastern has benefited from taking a framework that our predecessors established a century ago as a basic applied learning approach and evolving it into a global experiential learning model that provides a unique level of breadth and depth for learners. As they make their way into their careers and the world, our students profit from having used co-op to find out what they like and do not like, to discover what they are good at and what they are not good at, and to understand the diversity of the human family and the human experience in meaningful ways. At the same time, all colleges and universities have the opportunity to leverage elements of experiential learning, particularly in the residential campus setting. After all, a campus is simply another environment to experience, and it is a particularly rich one that pools a concentration of resources and people.

First and most obviously, the residential college model inherently cultivates cultural agility, drawing together students from diverse backgrounds and philosophies. When disparate people see each other every day in class or in the dormitory, they have no choice but to observe their commonalities and differences. Years ago, two freshmen, a Muslim student and a Jewish student, were assigned to be roommates and immediately demanded (but did not get) alternate lodging partners. By the Thanksgiving break, they were celebrating the holiday in each other's homes. Because many undergraduates have grown up with limited exposure to other cultures, the educational importance of a diverse student body cannot be overstated.

Although the foremost purpose of the college is to educate students through challenging coursework, students can hone their critical and systems thinking in many other ways. Clubs and student organizations provide goals, projects, and experience with leadership and teamwork. Volunteering and service learning opportunities do similarly, with the additional force of having real consequences. For example, a student joins the campus chapter of a national mentorship organization in her freshman year, thereby developing her cultural agility but also her human literacies and communication skills. During her sophomore year, she further hones her human literacy as a writer for a student magazine, then practices her critical and systems thinking as an outreach coordinator for Black History Month, connecting people from the campus and the surrounding community. In her junior year, she takes a yoga teacher-training class, which she parlays into organizing and leading a yoga retreat to South America—a great example of the intersection of cultural agility and entrepreneurship. Learning happens everywhere. The salient point is for students to be cognizant of and hence reflective about their experiences.

Entrepreneurship experiences do this especially well because they yield very measurable results. Given the right conditions, entrepreneurship can be fully woven into the fabric of campus life, greatly expanding its educational reach. One study showed that, within the workplace, peers influence each other to spot opportunities and act on them: the more entrepreneurs you have working together in an office, the more likely their colleagues will catch the bug.[20] A study of Stanford University alumni found that those "who have varied work and educational backgrounds are much more likely to start their own businesses than those who have focused on one role at work or concentrated in one subject at school."[21]

To cultivate an entrepreneurial culture, colleges and universities need to offer students a broad choice of experiences and wide exposure to different ideas. They are uniquely positioned to do this by combining the resources of academic programming, residential life, student groups, and alumni networks. Student-run venture incubators, for example, can take students through the comprehensive process of launching startups—from idea generation through business planning, development, investment, and implementation.

Consider a young engineer and entrepreneur who has an idea for a special coating that protects implantable medical devices—and the people who have them—by repelling bacteria. She takes it to the venture incubator, where her peers studying biochemistry help her assess the biological interactions the coating would have with the device and with human tissue. Colleagues from the business school could help her research the market for her product and roll out a business plan, while design majors might help her develop the branding for the product. She could also connect with alumni mentors from the health sciences and venture capital communities who would offer advice on product launch, leadership, or even financial support. Meanwhile, faculty provide oversight and advice from an academic perspective.

Because faculty members already assist students in exploring original research opportunities beyond the usual curricular requirements, they can do this in a way that promotes reflection, iteration, and integration with coursework. By generating new knowledge in contexts outside the classroom, students are creatively energized and tested on the mettle of their critical and systems thinking. The undergraduate whose name appears on a piece of original research may furthermore make discovery a fulfilling, robot-proof career path.

Global experiences, too, take advantage of the educational richness that comes with a radical shift in the student's learning context. By stepping away from the controlled setting of the campus into the maelstrom of a broader reality, the student's experiential learning goes into overdrive. Just as important, sustained interaction and observation of people from different cultures teaches cultural agility. It gives students culture-specific knowledge as well as practice in reading contexts, weighing responses, and learning how best to integrate, adapt, or tactfully deflect a situation. For example, a U.S. college student who spends six months working for a New York bank may learn about international finance, but if she works for a bank in Hong Kong, she also will learn to negotiate the unfamiliar terrain of Chinese office politics, grocery shopping in Cantonese, and riding the Star Ferry to Kowloon. Swimming deep in these unfamiliar contexts, she exercises her mental flexibility and creative problem-solving.

THE EXPERIENTIAL LIBERAL ARTS[22]

With all the possibilities offered by co-curricular experiences, most colleges will have no difficulties in incorporating experiential learning into their programs. Yet liberal arts institutions may seek a more structured approach that explicitly integrates the arts and humanities with real-life experience. Called "experiential liberal arts," this model advances past the old juxtaposition of the humanities versus the applied disciplines, tapping into the complexity and rigor of both.[23]

The liberal arts are, regrettably, all too often dismissed as impractical—as a student's quickest route to her parents' basement by way of an anthropology degree. Nothing could be more mistaken. The study of applied disciplines such as engineering sets a steep academic challenge, but so does the study of any complex system. Engineers must learn the dizzying intricacies of a structure's composition and context, including materials, environment, the forces of physics, time, and logic. Such complexity spurs intellectual development, yet the study of human culture and behavior is just as complex. Both demand the grasp of intricate systems, with the humanities and social sciences offering an elaborate mesh of history, art, geography, economics.

For instance, when an engineer fills her automobile with gasoline, she might muse about the workings of her car's engine or about the process of oil extraction and refinement. A person educated in the liberal arts, on other hand, might reflect on the history of OPEC, international energy use, and global warming. Recalling her philosophy classes, she might think it ethically necessary to drive less or consider the role of oil in fueling international conflicts. In regards to rigor, then, the liberal arts can be just as demanding as any "harder" disciplines.

Perhaps the reason that applied studies and hard science often possess a reputation for greater utility is that they feature a strong laboratory or workplace component while liberal arts classes are often framed by written self-expression. There is an arguably arbitrary distinction between the abstract and the real, but it may account for the conceit that a liberal arts education does not lead to a solid paycheck. However, by reproducing the lessons from engineering laboratories or business school internships, liberal arts programs can counter this misperception. They need an experiential component. This means combining the rigor of traditional academics with active participation in workplaces, laboratories, or volunteer opportunities.

For example, an English major might intern with a media company, applying ideas she encountered in a class on the technology of text to writing in new publishing formats. A philosophy major could parlay a work-study program at the United Nations Human Rights Council into a research project on labor ethics in the global economy. A history major could apply content knowledge and research skills to write a report analyzing plans for a proposed civic construction project. I know one undergraduate who applied his English major to designing a customer support system and performing financial analysis for a cloud computing startup firm. "Doing financial analysis is surprisingly similar to doing literary analysis," he told me. "When you read a poem or a novel, your professor tells you to look between the text and dig as deep as you can to find out everything the author is trying to say. When you're looking at a spreadsheet of numbers, you're doing the same thing: What are these numbers trying to tell me?"

An experiential liberal arts model also integrates traditional liberal arts skills with technological proficiencies. This gives students the tools of the digital humanities and computational social sciences, teaching them to apply data and technological literacies to human literacy, but it also pushes students to explore the social dimensions of our machines, including the ethical implications of technological change. Advances in computational analytics have revolutionized the humanities, so students should be prepared to utilize these technologies in full. They can now make connections that had, until recently, been impossible. For instance, a recent study out of the University of California at Santa Barbara, "Network Science on Belief System Dynamics under Logic Constraints," used a mathematical model to examine how people's shared belief systems are interlocked and affected by interpersonal influences.[24] Using tools such as this, humanities and social science scholars can untangle extremely knotty questions with unprecedented accuracy.

Experiential liberal arts students can use these new technological methods to work on original research, engage in student-faculty collaborations, and integrate learning and outreach through projects based in their communities. History students, for example, could compile digital archives of nineteenth-century black intellectuals in their cities or create networks of historical texts and maps. The humanities are expanding their digital toolbox. As we expand the scope of what they can do, we have to expand the scope of what we must teach.

The marriage of liberal arts skills with experiential learning and technological proficiencies is an ideal method for growing students' cognitive capacities and their appetites for ongoing study in new learning situations throughout their lives. Bates College, a leader in U.S. liberal arts education, is tapping into this method through its strategic priorities of "engaged liberal arts" and "purposeful work." Not long ago, Bates resolved to orient its curriculum toward helping "students identify and cultivate their interests and strengths and acquire the knowledge, experiences and relationships necessary to pursue their aspirations with imagination and integrity." This initiative involves skills-specific courses, as well as an emphasis on building workplace skills through co-curricular programs, paid internships, and practitioner-taught courses. In the college's intensive, short Spring Term, such courses include practical titles like "Business of the Arts," "Project Management," or "Brand Culture Building."[25] This embrace of workplace realities hasn't betrayed the ideals of this liberal arts college, but rather helps students develop their critical thinking and creativity in novel ways, while the stress on internships brings them to flower in the hotbed of the real world.

But experiential liberal arts experiences should not be only for liberal arts students. Even brilliant computer scientists have to thrive in a human context or risk having their work overlooked. Even trailblazing biochemists need to understand the social implications of their research. By enhancing their courses of study with the humanities and social sciences, programs in the "harder" disciplines can better prepare their graduates to succeed. By bleeding a little into each other, both approaches to higher education give graduates a powerful, practical education.

ASSESSING EXPERIENCE

Whether it is through co-op, co-curricular experiences or the experiential liberal arts, experiential learning in higher education aims to level the walls between the classroom and the rest of life. Learning happens everywhere, so educators must help students grow from their experiences wherever they occur. To that end, we need to consider how students can turn implicit, even unconscious learning opportunities into explicit learning outcomes.

One way to do this is to map students' growth as they move through the learning ecosystem. This would mean that classroom grades are only one facet of their developmental assessment. Educators also would track students' advancement in their new literacies and cognitive capacities, wherever they practice them. A student who volunteers as a mentor for disadvantaged girls, for example, would significantly boost her human literacy and cultural agility. If she publishes articles in a campus magazine, she would improve her critical thinking and creativity. Throughout the entire run of her time in college, each class, activity, and experience would add to the tally of her overall development.

So much learning occurs in the particulars of daily existence that it may seem futile to attempt to quantify them. Educators can log students' co-ops, trips abroad, and even yoga classes, but it is hard to envision an assessment of student learning throughout the entire learning ecosystem. What about the inspiring political event they attend at the student center? Or the debate on business and environmental regulation that erupts, after pizza, in a dorm's common room? These, too, are microlearning experiences, and students can benefit from them even more if they are cognizant of their learning as it occurs, remembering to reflect on the lessons afterward.

Educators are not going to shadow students around the clock, whispering words of mindfulness. But universities can use technology to equip students with tools to make them aware of the process of continual learning. At the university I lead, students can use their devices to tap into an app called SAIL (Student Assessed Integrated Learning), in which they record learning experiences—including serendipitous ones—and track their progress in different developmental dimensions and skills.[26]

For example, a student who auditions for an improv troupe would record that she practiced creativity, teamwork, and communication, and she could see her progress visualized in a clear, intelligible graphic. She might even use the app to upload pictures of her experiences and share her self-reflections, articulating what she learned and how. Tapping into the techniques of gamification and social media, tools such as this can help us better personalize students' learning and give them greater control of their own development.

The greatest teacher is life itself. Through innovative approaches, universities can find new ways to draw on that power. Then our graduates will be ready for all of life's challenges.

EXPERIENTIAL LEARNING FOR LIFE

We have explored how a humanics education catalyzed by experiential learning is the surest route to a robot-proof future. Yet there remains the outlying fact with which this book began: computers, software, and artificial intelligence are getting better on an exponential curve. The machines are marching on.

Much as our devices require periodic software updates, so do our biological brains. A fundamental consequence of technological progress is that people need to shift their belief that higher education is something that takes place at only one or two pivotal times in life to an understanding that it is a process that is lifelong.

Some learners are long on time and short on experience, so we can reach them through the methods articulated in this chapter. Other learners are short on time but long on experience, so we have to create learning opportunities within their existing careers and institutions. The realities of our time demand that we redefine our notion of what it means to be a student and an alumnus. To that end, universities must build on the existing architecture of higher education, creating a new model for lifelong learning beyond the scope of today's online classes and after-work programs.

Higher education, like all of us, will have to adapt.

LEARNING FOR LIFE

In *A Study of History*, the British historian Arnold Toynbee argues that civilizations, like individuals, thrive when they successfully answer challenges with creative responses. Faced with a growing population and limited means of subsistence, for example, ancient Athens avoided revolution by developing mercantile trade and democratic institutions.[1] But the wheels of progress slip from their axles when a civilization stops responding to new challenges in a creative way. After struggling in the aftermath of the Peloponnesian War, Athenian democracy could find no answer to the rise of Macedon, and Athens ceased to be an independent power. Without innovation, a society slumps into decay and eventually ruin.

Toynbee's argument is that decline is essentially a cultural failure—a willful stagnancy born from what Carol Dweck might call a "fixed mindset" but that Toynbee characterized as stifled creativity. History is littered with the fossils of societies that, as the dinosaurs or the dodo, were unable to adapt to circumstance. Carrying Toynbee's admonitions further, historian Jared Diamond describes how societies such as those on Easter Island and the Viking settlements in Greenland clung to cultural habits despite creeping change—in these cases, the self-inflicted ecological wounds of deforestation and soil erosion. Despite the visible encroach of doom—spoiled landscapes, hungry winters—they refused to change their ways and let themselves sink into the archeological record.[2]

Modern colleges and universities are among the fullest expressions of human culture that have ever flowered. As fertile grounds for cultivation of the mind, they are unprecedented; as hothouses for knowledge, they have no equal. They are perhaps the most effective institutions for intellectual advancement ever developed by humanity. Nonetheless, if colleges and universities fail to respond creatively to the challenges they face, they, too, will wither into irrelevance.

The university today is an ideal engine for delivering a standardized form of higher education in a defined period of time. It is a mechanism structured for offering deep but often depersonalized access to knowledge. Its culture is shaped by the goals and forms of that mechanism, serving the components of disciplinary departments, degrees, and faculties. It is very good at what it does. The problem is that in the twenty-first century we need it to do more.

In the previous chapters, I argue that advanced machines are revolutionizing the global economy and hence are poised to disrupt every other aspect of society. If our avenues for work, wealth, and well-being fork in new directions, then education, by extension, must follow. Therefore, universities have the opportunity to respond to changing realities by evolving their curricula, yes, but also by adapting the mechanisms for advancing those curricula. They have a chance to update their structural components. In other words, we can strip the engine down to its frame and rebuild it.

In addition to serving students as a group, higher education can be refit to serve the individual learner in a personalized and customized way. As is shown in previous chapters, this is a response to the intricacy of the AI economy and the demands it will make of human professionals. Moreover, instead of serving people during isolated fragments of time (four years in early adulthood and perhaps additional years in midcareer), higher education increasingly will be compelled to serve people throughout careers defined by continuous technological change. To do this, the university must be rethought in all its parts, bringing lifelong learning to its center.

As machines leapfrog in their abilities, they will continue to eliminate entire categories of white-collar, knowledge economy jobs. At the same time, technology also will bring about new jobs and new industries that will require people to acquire more advanced knowledge and skills. Hence, as machines advance, *all* people will need to retool, refresh, and advance their knowledge and skill sets on an ongoing basis. The logical conclusion is that to stay relevant in the AI economy, lifelong learning will be an imperative for all professionals—and not only professionals. By helping everyone develop and maintain valuable skills, lifelong learning is necessary to alleviate social inequality. A learning model oriented on that goal will serve both those who are long on time but short on experience (namely, recent graduates) as well as learners who are short on time but long on experience (namely, seasoned professionals). Consequently, colleges and universities will see benefits in making lifelong learning a focal point of what they do.

THE HUMBLE BEGINNINGS OF LIFELONG LEARNING

In many ways, the rise of the lifelong learning imperative today is a case of past as prologue. In a previous era of intense technological change—the years leading up to and during the Industrial Revolution—most people lacked the opportunity to earn a college degree. However, they responded to changing times and working environments by acquiring various types of lifelong learning.

For example, in 1831, the year Charles Darwin set sail on the *Beagle* for his famous encounter with the fauna of South America, another young Englishman named Isaac Pitman began earning his certification as a teacher. A lifelong educator, Pitman, like his better-known contemporary, had particular views about the relationship between life and time. Whereas Darwin thought about natural selection over generations, Pitman liked to say that "time saved is life gained"—a creed he served by inventing a system for taking phonetic shorthand notes. Known as "Pitman shorthand," it became the most popular language for transcription around the English-speaking world. It still remains the most widely used phonetic system in the United Kingdom.

The Pitman system's rapid spread and popularity was no accident. To reach potential students who did not live near his home in the city of Bath, Pitman took advantage of a newly invented technology—the standard postage stamp. He mailed exercises to subscribers and then corrected the returned postcards to provide them feedback.[3] This long-range format of instruction and response was the world's first distance learning course.

Ever since Pitman first tapped into a cheap and reliable postal network, distance learning has been a staple of lifelong learning, although it is by no means the only way in which to deliver it. In early nineteenth-century London and Boston, for instance, learned gentlemen banded together to found the Society for the Diffusion

of Useful Knowledge, which offered educational lectures and pub-
lications to the working public. The Boston group hosted talks by
thinkers ranging from Daniel Webster on "The Progress of Popular
Knowledge" to Oliver Wendell Holmes on "Homeopathy, and Its
Kindred Delusions."[4] In 1836, Boston philanthropist John Lowell
Jr. left a large bequest to pay for edifying public lectures. This was
the basis of the Lowell Institute, which, over the decades, has given
rise to educational institutions ranging from public broadcasting
stations to Harvard's extension school and the Lowell Institute
School at my own university.[5] To this day, the Lowell Institute helps
learners with a partial college education complete their degrees. It
is a further chapter in a long tradition of tackling inequality and
straitened circumstances through lifelong learning.

Early lifelong learning formats also grew out of a mission
for moral, as well as more worldly, instruction. In 1844, George
Williams—a displaced farmer learning to navigate the perils and
temptations of modern London—founded the first Young Men's
Christian Association (YMCA) to promote Bible study and social
cohesion in the swelling ranks of the new urbanites.[6] A few years
later, Thomas Valentine Sullivan brought the institution to Boston.
Soon, YMCAs were teaching immigrants English as a second
language, helping them ease the dislocation of abandoning their
native lands for the tumult and change of the New World, and
offering vocational classes to working people dreaming of higher
pay and better conditions. In all these cases, lifelong learning was
a way for already experienced workers to adapt to their evolving
circumstances.

Darwin and Pitman lived in a time of extraordinary technological
and social change, and both of their careers were deeply informed

by ideas about the nature of progress. On his voyage, Darwin began to formulate his theory of natural selection by observing variations in how species adapt to their environmental contexts, passing down traits that promote survival. For instance, he noticed that some Galapagos finches had differently shaped beaks that gave them an advantage at acquiring particular foods. Meanwhile, as Darwin pondered his theories, Pitman's correspondence pupils were learning a valuable office skill, enabling them to survive better in the quickly expanding environment of the modern, industrial cities of the British empire and the United States. In a competitive system, education was an equalizer.

But although finches can pass down the genes that give them long beaks to drill into the fleshy parts of a prickly pear, human beings cannot pass down a gene for writing shorthand. Instead, we evolve by getting ourselves educated. That is why as we take stock of what many are calling the Fourth Industrial Revolution, lifelong learning is again as important as it was during the last one.

THE DEMAND FOR LIFELONG LEARNING

Higher education has ample experience serving older and non-traditional learners. Of the 20.5 million students attending U.S. colleges and universities in 2016, 8.2 million were twenty-five years or older.[7] A full 40 percent of students, therefore, are older than the age generally viewed as "traditional" for college. By 2025, the number of students aged twenty-five or older is projected to increase to 9.7 million.[8]

Many of these students are served by the more than one thousand community colleges in the United States.[9] For generations, these have been the standard-bearers for extending the promise of higher education to the most vulnerable and underserved populations in U.S. society, including lifelong learners such as the employee whose factory has closed, the recent immigrant from a country with a less-developed educational system, or the single parent who did not complete high school. The ranks of the vulnerable now include those threatened by technological change. This renders the mission of community colleges even more vital—a mission that they traditionally have met by giving students a conduit to a four-year degree and teaching career-oriented skills. The demand is clearly great: some 12.8 million students attend them annually.

Thus, it is somewhat surprising that other types of colleges and universities often consider lifelong learning to be peripheral to their mission. Today, it is not uncommon for institutions of higher education to have some kind of organized enterprise for lifelong learning, such as an extension school, a college of professional studies, or a continuing education division. All too often, however, they are relegated to secondary, even second-class status. Educating undergraduates, preparing graduate students, and creating new knowledge by conducting research are seen as the real, serious endeavors of the university, while lifelong learning is viewed as ancillary.

Undergraduate education, graduate education, and research are indeed critical for core priorities. But the traditional approaches will not work for the millions of adult learners finding themselves compelled to return to higher education to stay ahead of technological change. In the past, universities adopted a *Field of Dreams* sort of approach to their enterprise: build it, and they will come, they said. As such, they constructed departments and programs offering the expected stable of traditional, monolithic degrees such as the bachelor's, the professional master's, and the PhD. This is no longer enough.

By choosing not to adapt to—and prioritize—the needs of life-long learners, colleges and universities are overlooking a vital educational need, especially in our hypertechnological reality. In many ways, their attitude is reminiscent of the posture that railroad companies took during the early decades of the twentieth century, when airplanes first came on the commercial scene. At the time, the railroads had cornered the market for long-distance passenger transportation. When the first airlines also began to offer long-range commercial flights, the railroads largely ignored this development, considering air transportation to be a fundamentally different endeavor from the business in which they were occupied. They saw the airlines as being in the airplane business, while they saw themselves, naturally, as being in the train business.

The railroads failed to realize that both they and the airlines were actually in the same business—namely, transportation. Consequently, they missed the warning signs pointing to the impending disruption of their industry. And when commercial air travel took off, as it were, they were shocked to see their new competitors quickly dismantle their longstanding dominance in passenger transport.

As AI, robotics, and high technology give rise to an unprecedented need for people to learn, retool, and upskill throughout their lives, higher education would do well to consider shifting its perspective in the way the railroads failed to do. Going forward, colleges and universities have the chance to recognize that they are not merely in the specific businesses of undergraduate education, graduate education, and research—although all of those remain vitally important. Rather, they are in the larger business of lifelong learning.

As a matter of fact, precisely because the higher education sector largely has yet to shift its perspective in this way, others—most notably, for-profit colleges—have stepped in to fill the breach. Between 1990 and 2010, enrollment in for-profit colleges boomed in the United States and around the world. Much of this demand came from older students and working professionals who were attracted by the flexibility of the online model used by most for-profits. In academic year 2007–2008, for example, only 11 percent of students enrolled in for-profit colleges were the "traditional" college ages of eighteen to twenty-three.[10] And although for-profit college enrollment has since receded as some have become embroiled in scandals over allegations that they overstated their graduates' job placement rates, the overall trend clearly shows that the appetite for lifelong learning in the market is strong.[11]

It is not only for-profits that have taken up the banner of lifelong learning. There also has been an upsurge in "corporate universities," or in-house academies for training managers. General Electric is credited with pioneering the approach in the 1950s, and the model has exploded in recent decades.[12] Boston Consulting Group estimates that their number doubled between 1997 and 2007, recently reaching about five thousand corporate universities worldwide.[13] As a corollary to this model, some companies are partnering with nontraditional providers to offer further education to their employees. For example, AT&T is working with MOOC (massive open online course) provider Udacity to offer its employees the chance to upskill—and giving them negative performance reviews if they choose not to invest their own time in taking courses.[14]

The "corporate university" model has many appealing facets. It allows companies to tailor employees' learning to their particular business needs. It also can serve as a pipeline for training managers within the firm's culture. At the same time, it fails to account for one of Warren Buffett's basic investment tenets: stick to what you know. Very few enterprises besides colleges and universities are in the business of higher education. Thus, when companies set up in-house education programs, they are not playing to their strengths.

The rise of in-house corporate education is further evidence that higher education is sidelining lifelong learning to its detriment. Education is what colleges and universities do best, so companies should not have to take up the academic mantle. It makes better business sense to partner with the experts. If my university is interested in selling clothing in our school colors, instead of building our own garment factory, we outsource the job to an established clothing manufacturer. The very fact that for-profits and corporate universities have seen such growth shows that higher education is failing to serve its natural constituencies.

This missed opportunity is especially unfortunate because today's professionals are facing challenges as profound as those faced by the workers of Pitman's and Darwin's day. Just like them, they are immersed in rapidly changing work environments to which they must adapt or risk losing their competitiveness. Just like them, the escalation of technology means that they must increase their uniquely human skills through further education. And just like them, in an increasingly complex economy, lifelong learning may well be the difference between their professional evolution and their economic extinction. In this context, the old *Field of Dreams* approach no longer suffices: universities cannot simply build monolithic programs and expect lifelong learners to show up. Instead, effective programs will have to be customized and personalized for the growing cadre of lifelong learners.

A CUSTOMIZED, PERSONALIZED MODEL

Customized and personalized lifelong learning begins with a simple admission. It accepts that as technological change drives workplace realities, higher education has an obligation, and an opportunity, to respond. This simple truth distinguishes it from the "build it and they will come" model, which may or may not be swayed by factual changes in the real world. The old model is university-centric, with course design often dictated from on high, often predicated on assumptions about the workplace that may be obsolete by the time the ink in the textbooks is dry. That approach will not work for learners stepping into complex, evolving roles in the AI economy. To adapt to their needs, we might consider how customization is relevant to lifelong learning in two dimensions— design and delivery.

Customization and Design

In the past, universities determined curricular design largely by themselves, assuming that they understood learners' needs and desired outcomes. In the lifelong learning model of the future, universities will codesign curricula in full partnership with employers and learners. From a conceptual point of view, this means they will have to sit down with learners to map out their professional needs and outcomes candidly. Just as important, they will respond to changes in the workplace by inviting employers to discuss their business demands. This means that in the conceptualization of new academic programs, universities will account for the business strategies employers are seeking to accomplish, their professional workforce requirements, and the ways technology is changing the shape of their industry. In other words, the employer will become an equal partner in delineating the contours of educational content, helping to keep it streamlined and relevant to the moment.

We can see some good examples of this dynamic at work at the undergraduate level, but the lessons can apply to lifelong learning programs. For instance, the University System of Maryland is spearheading a multipartner collaboration between higher education institutions, businesses, and government agencies to address the state's and the nation's cybersecurity workforce challenges.[15] This includes numerous initiatives between employers and universities, such as the Advanced Cybersecurity Experiences for Students program developed between Northrop Grumman and the University of Maryland at College Park. Designed to educate cybersecurity professionals and future leaders, the program is grounded in honors-level academic courses that have been codeveloped by industry representatives and complemented by cocurricular experiences and real world project-based learning. In a similar mode, Illinois State University has partnered with State Farm, the large insurance company, to develop more robust cybersecurity programs, including sponsoring cybersecurity "hacker challenges" to foment student interest.[16]

These sorts of partnerships have emerged in broader contexts as well. Not long ago, IBM partnered with a number of universities, including Carnegie Mellon and the University of California at Berkeley, to design cognitive computing courses using the Watson technology.[17] The goal of the courses is to meet the booming demand for data analytics professionals by embedding IBM's technology in the classroom and pairing students with the company's "technical mentors." In this way, learners are educated in the industry's latest concepts and tools well before their first day on the job.

Another iteration on university-employer collaboration is Northeastern's ALIGN program, based at our Seattle campus. This initiative is explicitly designed to funnel people from diverse backgrounds into technology careers such as bioinformatics or cybersecurity. Exploring opportunities on the West Coast, we found that many potential students possessed bachelor's degrees that did not match with local job openings. They wanted on-ramps into high-demand technology fields driven by local employers. Thus, we worked with employers to develop a program to allow liberal arts graduates to become computer scientists, providing them with a master's degree and up to twelve months of co-op or internship experience in the tech sector.

The result combines intense curricular work with immersion in the workplace. Learners gain new high-tech skills—for example, in big-data analytics—by tackling content that is chunked out into on-the-job projects. These projects are coupled with online learning units to help learners master the latest technologies. Further, the program leverages learners' experiences in the liberal arts, integrating these with their newfound talents, giving them an edge on communication and critical thinking—crucial skills for management positions. At the end of the program, their new knowledge is bolstered by the competencies they have acquired in the past, prepping them for success.

Lifelong learning curricula need employer input for content but also for other elements. For instance, it is not much use if educators concoct a brilliant, transformative curriculum that clashes with workers' time commitments on the job. Employers and lifelong learners need to agree on working hours invested in education, much as they need to decide on appropriate rewards and incentives for employees who are lifelong learners, including promotions and financial support. Also, universities and groups of employers within an industry may even consider providing credentials jointly so that learners can move seamlessly between different employers in their field and have their education, knowledge, and skills appropriately recognized.

Customization and Delivery

In addition to customized design, the lifelong learning imperative driven by technological change also will require higher education delivery to be customized. In the past generation, online programs have been higher education's main answer to learners' need for a customized delivery format, and as a general strategy to assist learners who need flexibility and are short on time in *how* they receive an education, they are serviceable. They deliver learning wherever and whenever learners need it. However, from the standpoint of *what* they deliver in terms of knowledge and skills, many online programs do not go far enough. Often, they suffer from being overly generic. They typically are designed for large cohorts of learners, without necessarily taking the specific learning needs of people in specific industries into account.

Although studies vary, some indicate that purely online learning is less effective than a blended, hybrid mix of online and face-to-face components. For example, one meta-analysis from Stanford University found that on average, students performed about equally when receiving purely online instruction or in-person instruction, but those receiving hybrid learning performed best.[8] This may be because hybrid programs tend to involve "additional learning time, instructional resources, and course elements that encourage interaction among learners."[9] People are, after all, social animals. Hybrid formats possess the dual advantages of greater engagement through personal interaction and greater customization through online options.

But where both typical online and hybrid formats fall short is that their value is ultimately limited unless they include experiential components. Truly transformative learning results are the product of integrating academic and real world experiences. Yet the question of experiential components in lifelong learning also raises something of a paradox. Experience is vital for younger learners, but if learners are long on experience already, why do they need more learning experiences?

The answer is that in an AI economy, older learners will need to capitalize on the valuable lessons they have acquired in work and in life. By revisiting past experiences and using them as a compass to recalibrate their actions and ideas, they better position themselves to achieve their goals and create new opportunities and experiences. Lifelong learning programs that do this well will meet learners where they are—including in the professional workplace.

Such programs are especially useful in serving the people who need lifelong learning the most—the enormous subset of current white-collar employees who have always considered themselves secure from automation but who now suddenly hear the tramp of robot feet. This format embeds the education within employees' work itself, bringing lifelong learning directly to learners at the ground zero of technological upheaval. We can reach them where they see automation firsthand, while seamlessly integrating education with learners' existing experience.

As an example of this in action, Northeastern, with support from the U.S. Department of Education's Education Quality through Innovative Partnerships (EQUIP) program, collaborated with General Electric to develop an accelerated degree in advanced manufacturing. This is a customized, in-house program for upskilling GE employees to work alongside the next generation of technologies, training them for the jobs of the future even as they perform their jobs today. It operates on a project-based format that capitalizes on employees' experience and utilizes online components so they can learn on a "short on time" schedule. Students still perform their full-time jobs, but they simultaneously learn to undertake work in critical areas necessitating advanced training—in other words, robot-proof work.

While customization means designing programs to serve learners' and employers' needs jointly, personalization means molding the learning experience to students' strengths, weaknesses, ambitions, and schedules. For example, in the last chapter we encountered personalization through SAIL, the gamified technology for tracking students' experiential learning. Similarly, highly personalized career services and professional guidance help keep learners on course to achieve their educational goals.

Personalization also makes lifelong learning more accessible to gig economy freelancers, who face the same challenges as full-time employees but without the supportive infrastructure. A recent study from the Brookings Institution found that the number of these freelancers is increasing even more dramatically than generally believed.[20] In the past twenty years, according to one calculation, the number of people engaged in the gig economy grew by 27 percent more than growth in payroll employees.[21] Yet just like everyone else, freelancers face a labor market in which jobs increasingly will disappear and emerge with startling speed. Just like everyone else, they are witnessing the impact of AI and automation. And just like everyone else, freelancers in the gig economy can benefit from a learning model powered by experience.

By embedding in a variety of companies through lifelong learning programs, enterprising freelancers can learn different roles, ascertain talent gaps, spy opportunities for future work, and grow their personal networks. They can establish relationships with different hiring managers and gain a fluency in different company cultures. They may even set themselves up for a full-time position because many companies seek to "test drive" potential employees before committing to the investment of a long contract. As the number of gig economy freelancers grows, companies will increasingly rely on them for labor. Lifelong learning programs give both parties an opportunity to determine whether there is an amicable alignment.

SOME IMPLICATIONS FOR THE UNIVERSITY

Degrees and Credentials

Lifelong learners, by necessity, typically approach education in some different ways from the traditional full-time student. They usually do not have the luxury of committing to a fully articulated academic degree program, much as they might like to. Instead, they need an educational experience that is more focused and tactical. Often, lifelong learners are looking to acquire a targeted set of knowledge, skills, or competencies in order to meet a specific goal in their professional employment or the larger arc of their careers and their lives. They want to acquire this knowledge efficiently and operationalize it quickly and effectively.

As an example of this, I once advised an entrepreneur who was building a new type of Internet search engine and wanted to understand the contributions of linguistics (my field) to that endeavor. He did not want a degree: all he cared about was the knowledge itself. He needed a crash course in the application of linguistics to search technology, tailored to his precise goal. To do this, he needed to learn some fundamental knowledge about syntax, phonology, and semantics. At the same time, he also needed to grasp some of the much more advanced computational aspects of linguistics simultaneously. Pursuing a traditional university degree in linguistics would not have helped him achieve his goal of building a search engine. But I was able to help him—and so would a university that could nimbly curate and tailor the knowledge he needed.

The rising imperative to meet the needs of people like my entrepreneur colleague is one reason behind the growth of "boot camps" for lifelong learners offered by newly established players such as

General Assembly, Bit Bootcamp, and Data Science Dojo, as well as by traditional universities. For example, Northeastern's Level program is a data analytics boot camp designed for lifelong learners. Instead of enrolling in a lengthy master's program in that subject, students can sign up for a burst of focused content. Level's design is customized for professional outcomes: it is structured to align with job openings in growing fields and places learners in experiential projects with potential employers. Delivered in less time than a typical college semester, the program is also remarkably efficient.

Innovative program forms such as boot camps are helpful, but higher education does not need to stop there. The burgeoning demand for lifelong learning—and the acceleration of this demand as people seek to upskill in the AI age—suggests that we reimagine the larger issue of how colleges and universities organize knowledge, as well as how we parcel it out.

Today, colleges and universities erect artificial divides between knowledge taught to undergraduates and knowledge included in graduate-level curricula. As the population of lifelong learners grows and as colleges and universities focus more intently on the needs of these learners, we will need additional ways to organize and segment knowledge. Instead of fitting it into old categories such as undergraduate and graduate courses, we can chunk it into smaller modular blocks that can be assembled according to learners' targeted objectives. These blocks can then be combined and stacked in a way that is akin to a traditional degree, except that there will be many more permutations and combinations of these blocks than is possible in a typical degree program. Moreover, each possible combination will be much more tightly coupled to the specific needs of the learner than a typical degree allows.

For example, consider three learners who wish to study bioengineering. The first is a career changer who is seeking to break into the life sciences from a position in the technology field. Thus, the sequence of curricular blocks she takes might include four blocks of basic bioengineering content, combined with three advanced blocks that leverage her existing knowledge of technology. The second learner already works in bioengineering but is seeking to move up from a midlevel position at her company to the managerial ranks. In this case, the sequence of blocks she takes might include three blocks of bioengineering content to get her up to speed on the latest advances in her field combined with two blocks that focus on managerial competencies and business development. Finally, consider the third learner—a senior manager at a health care company that is seeking to expand the uses of health-related nanotechnology in the delivery of health care. In this case, the manager might take two blocks focused on health-related nanotechnology applications and nothing more.

The tight coupling of these educational sequences with these learners' needs is self-evident, but what really is novel is how the university might translate these stacked curriculum blocks into formal academic credentials. For example, we can imagine a university awarding the first learner a bachelor's degree in bioengineering, in recognition of the longer sequence of blocks she has taken as well as the more basic content she has studied. The second learner might be awarded a master's degree, seeing as how she has taken a briefer but more advanced series of blocks. Finally, the senior manager could be deemed as having earned a certificate, in view of the short but highly specialized series of blocks she has taken.

The possibilities are endless, but the point is straightforward. By segmenting knowledge in a more finely tuned way, universities can impart that knowledge in ways that are much more effective at meeting learners'—and especially lifelong learners'—targeted needs. They can emphasize immediately functional knowledge, as opposed to ancillary knowledge that a results-oriented learner may never use.

Faculty

Elevating lifelong learning in the university's mission carries profound implications for students but also for all the other members of the university community. Increased demand for lifelong learning necessarily means that there will be an increased demand for teaching. As such, it is reasonable to expect that there will be a need for colleges and universities to expand their teaching faculties going forward.

This is not a new trend. According to the National Center for Educational Statistics, of the approximately 1.5 million college and university employees engaged mainly in instruction, research, or public service, nearly 1.2 million are primarily instructors.[22] In comparison, employees who balanced teaching and research amounted to a little more than 270,000, while pure researchers counted for a mere 65,000.[23] Moreover, the number of university-level teachers is growing further. Between 2004 and 2010, total campus teaching staff in the United States grew by about 200,000.[24] Full-time instructors increased by about 11 percent, while part-time adjuncts increased their number by nearly 30 percent.[25] This growing demand for instructors is only going to intensify when large groups of lifelong learners join the classroom.

The forthcoming influx of teaching faculty may also have consequences for the way in which faculty members do (and sometimes do not) work together. Nowadays, university faculties are tiered, with tenured professors who are mostly concerned with research occupying the upper echelon of the hierarchy. Outside this tier are nontenure-track faculty who typically specialize in teaching, professors of practice who come to the university to teach after distinguished careers in industry, as well as research fellows, adjunct professors, and more.

This system is called into question, however, in an institution premised on serving more lifelong learners, offering more stackable content, and puncturing disciplinary boundaries. Going forward, the vital role of the university teacher will be recognized more fully—and rightfully so. As more universities implement ways to deliver content in modular, stackable forms that are customized to learners' needs, teaching faculty across disciplines are likely to collaborate more often. Learners themselves will drive this change because they will increasingly ask their professors to bring diverse content blocks together in a way that meets their highly targeted needs, thus encouraging cross-disciplinary collaboration. Similarly, as universities expand their efforts to educate lifelong learners, it also is likely that faculty themselves will be encouraged to invest more time in their own lifelong learning by staying abreast of new program forms, delivery modes, and pedagogical techniques attuned to this population.

Alumni

Finally, augmenting the place of lifelong learning at the university will also test traditional notions of who alumni are and how alumni view their university. Most graduates of four-year higher education consider their alma mater to be the place where they received their bachelor's degree—whether or not they subsequently obtained a graduate or professional degree. Typically, alumni return to their college sporadically to attend sporting events and reunions or to tap into established relationships—in other words, to engage with the past. In contrast, the growth of lifelong learning at universities— and the need for more college-educated individuals to obtain it—can transform graduates' relationship with their alma mater, making them members of a widespread, active network engaged with the present and the future.

In addition to relying on their university for continuous learning, alumni will also stay connected through the ongoing use of career services and other support, with the operation evolving from a social capacity to a functional one. This is already happening in many universities, and it will soon become the norm. Thus, the standard alumni operation of the future will provide lifelong learners with access to venture incubators and startup assistance. It will connect professionals with accomplished mentors, offering coaching and institutional support. It will connect alumni businesses with faculty expertise and research. Furthermore, it will be the focal point for communities of interest, drawing together alumni who share professional goals, hobbies, or philanthropic objectives.

In this way, the literal meaning of *alma mater* ("nourishing mother") will be reinforced as graduates continue to be nourished and supported by their institution for their entire lives. In effect, they will become members of a lifelong club that fills enormous social, professional, and educational niches. This dynamic also has the potential to transform university giving. For example, in addition to writing an annual check to a class fundraising drive, alumni may enroll in a subscription model, signing up for continued access to services and learning opportunities. The entire advancement enterprise may evolve to rely less on alumni emotion and more on a manifest list of benefits, courses, and services rendered.

The expanded importance of alumni in a lifelong learning–oriented university raises a number of questions as well. As noted, alumni of U.S. colleges and universities tend to feel the most loyalty to their undergraduate alma maters. When people spend several of their most exuberant, youthful years living in a residence with their closest friends, they are likely to feel more attachment than if they sign up for a targeted chunk of learning delivered in a hybrid format. After all, students don't buy college sweatshirts to celebrate having completed a boot camp—at least not yet.

As people adopt a lifelong learning mode of education in the future, however, we may see traditional alumni loyalties change. People may have even more alma maters than they do today. The demarcation of gaining a standard undergraduate degree may give way to new credential forms. Consequently, alumni may end up feeling the strongest tug of loyalty to the university that has given them the most value over their lifelong careers. This is not necessarily a shift to a strictly transactional relationship, but when numerous institutions are competing for a person's limited attention, the standout will likely be the one that produces tangible results. If a university remains a dependable presence and support in their lives, learners are more likely to return to it for help, education, and enjoyment.

THE RISE OF THE MULTI-UNIVERSITY NETWORK

To this point, we have considered the internal implications for a college or university that seeks to elevate lifelong learning by shifting from a "build it and they will come" educational model to a model that is built around the learner. But the implications are not just internal. They also herald changes and opportunities for the overall structure of the university. It is no surprise that this should be the case. Throughout history, higher education ultimately has responded to changes in technology and to the landscape of learners it aims to serve by evolving its institutional forms and its overall shape as a sector. That is how, in previous eras, higher education in the United States came to encompass the agricultural and mechanical university, liberal arts college, research university, public university system, community college, and online university.

Today, we have once again reached such an inflection point. The oncoming wave of automation, AI-powered machines, and integration of high technology into all forms of work necessitates that virtually everyone become a lifelong learner—and that, in turn, necessitates that universities go to where the learners are. In the preceding section, I have outlined how this would be done in the figurative sense of aligning courses of study, faculties, and alumni experiences in ways that are learner-centered. But I mean this in literal sense as well. In an AI-driven economy in which the need for lifelong learning is ascendant and must be met, universities have the opportunity to go beyond existing as one institution or even a set of institutions within a state system. The opportunity is to take what I believe is the next step in higher education's evolution—the *multi-university network*.

Knowledgeable readers will note the similarity between this phrase and the term *multiversity* that Clark Kerr described in his classic volume *The Uses of the University*. But although the names are similar, the concepts are distinct. Kerr was describing the agglomeration of activities, interests, and people that comprised the U.S. research university as an institution in the 1960s.[26] In contrast, the *multi-university network* is a multilocation entity existing across multiple states and even multiple countries. Each node of the network is connected to the other, such that learners can circulate through it to take advantage of academic programs, learning resources, and experiential learning opportunities. In many ways, it is the next logical iteration of a university, taking into account the forthcoming need to serve a growing population of lifelong learners.

How Would a Multi-University Network Operate?

Consider, for example, a U.S.-based multi-university network with locations in Boston, Charlotte, Seattle, and Silicon Valley, and a young learner—a budding social media entrepreneur—whose future will be buffeted by the winds of AI-driven technological change. She might begin her lifelong learning trajectory by pursuing a customized series of modules in computer science and business at the Boston node of the network. She rounds out this experience with Seattle-based co-ops at Amazon and the law firm Perkins Coie, where she learns about intellectual property law. While there, she also takes an analytics bootcamp at the Seattle node of the network.

After utilizing this knowledge and experience to launch a new social media venture, she returns to the network a few years later when she predicts that her company, now successful, may soon founder because her customers are rapidly gravitating to virtual reality–based technology platforms rather than the traditional Internet platform on which her company relies. Consequently, she decides to enroll (along with key members of her company team) in a pair of modules offered at the Silicon Valley location that focus on how to transition businesses to VR-based platforms quickly and efficiently. Her gambit is successful, and her company continues to grow.

Five years later, as she sees even more technological changes emerging on the business horizon, she decides that the time is right to sell her company. So she returns to the network one more time—this time, to Charlotte, where she takes a refresher module on business acquisition and negotiation and encounters a new experiential learning opportunity that connects her to the area's financial institutions and potential financing for her next venture.

The Global Multi-University Network

Although the foregoing example is hypothetical, the reality of a true multi-university network is not far off. Indeed, Northeastern has been putting the fundamentals of such a network in place by establishing new, light-footprint campuses in the locations used in my example, as well as a non-U.S. location in Toronto. We are not alone: Carnegie Mellon has taken a similar approach in expanding its graduate degree programs throughout the world. Yale offers liberal arts education in Singapore, and New York University now boasts more than ten locations outside the United States. By adopting a multicampus, multimodal design, our institutions and others like ours will better meet the immense demand for lifelong learning that will present itself in the AI age.

Not every university will aspire to establish multiple locations in the U.S. or beyond—and not every university should. Linking disparate locations together to form a truly connected system is an enormously complex endeavor. Aligning academic programs and learning opportunities to allow learners to circulate through the network in a way that also is highly customized to their learning needs would require extraordinary coordination. Moreover, to serve lifelong learners most effectively, such a network ideally would also incorporate experiential learning and elements of the humanics model discussed earlier in this book.

Universities that seek to establish a multi-university network that crosses national boundaries will need to address an especially important dynamic—how to operate in manner that is truly *global* as opposed to merely international. Other industries have grappled with this dynamic. Years ago, when U.S. automobile companies began to expand outward, they adopted an international approach. They opened factories in new countries, but all decisions regarding products and strategy remained centralized at U.S. headquarters. This may have been efficient, but it was not particularly effective because the automakers (not surprisingly) often turned out new cars that were copies of the U.S. versions as opposed to distinct vehicles that captured the needs of customers in the new markets.

In contrast, when the pharmaceutical industry began exploring new frontiers, its leaders adopted a more global approach. They also built new facilities in untapped markets, but instead of simply replicating the strategies that had worked in the United States, they allowed their new operations to adapt to the immediate facts on the ground. Instead of pronouncing product design or marketing from on high, a decentralized approach prevailed that accounted for the opportunities and idiosyncrasies of place.[27]

Given the choice between going international and going global, I would argue that institutions of higher education seeking to establish a multi-university network would do well to follow the example of the pharmaceutical industry. By adapting to the realities of each locale in which they exist, such a network would embed an unparalleled degree of nuance in the education it provides. As learners circulate through the network, they would reap the rewards of this nuance—for instance, with deep cultural agility skills and extraordinarily nimble systems thinking abilities. By cycling them through different experiences and environments, a global multi-university network would foster in learners an outstanding appreciation for the diversity and variety of the world, enabling them to become creative and mentally flexible beyond the reach of machines.

Thus, consider the example of a lifelong learner whose twin passions are engineering and climate change and who enrolls as a student in a multi-university network with locations in New York, Vancouver, Dubai, and New Delhi. Circulating throughout this network, she could immerse herself in understanding urban coastal sustainability issues on the eastern U.S. seaboard and then pivot to the studying the questions engineers face in building effective water delivery systems in the Middle East. Pivoting once more, she might turn her attention to the understanding the health impacts of inadequate municipal sewer systems on the Indian subcontinent and then focus on assessing the impact of Vancouver's policies that promote renewable energy and green practices. The opportunity for this learner to learn this way, benefiting from customized programs and co-ops as she proceeds, would give her a truly unsurpassed education. Over time, she would obtain a more comprehensive—and complex—grasp of her subjects, gaining both a bird's-eye global perspective as well as a specialist's understanding of the subjects as they impact local communities. Once again, bringing this vision to full fruition would be no easy task. However, if an institution of higher education did so, it would produce robot-proof learners *par excellence.*

Clark Kerr's multiversity was a reflection of its world, just as the multi-university network is a reflection of ours. As our world doubles and redoubles in complexity, we cannot hope to understand it, affect it, and improve it except through commensurate complexity—in other words, through networks. To face the technological and economic challenges of the twenty-first-century successfully, individual learners will benefit from the opportunity to tap into global networks of higher education opportunities throughout their lives. To help learners seize this promise, many institutions of higher education, I believe, will inevitably evolve from their present forms.

AFTERWORD

Imagine the state of the world on November 17, 1944. The planet was home to some 2.5 billion human beings, none of whom had ever used a programmable digital electronic computer. The tide of war across Europe and the Pacific had shifted enough that in the United States, officials were thinking in earnest about how to reintegrate the millions of GIs who had been sent to fight back into the American economy and society.[1] Tuberculosis and premature birth were among the top ten causes of death among Americans.[2] Global surface temperatures were beginning their slow climb toward the troublingly high levels they would reach at the dawn of the twenty-first century.[3] The *Oxford English Dictionary* lacked an entry for "mobile phone," and even the one for "microchip" would not appear for several decades.

It was in that world, and on that day, that President Franklin D. Roosevelt sent a letter to Vannevar Bush, director of the U.S. Office of Scientific Research and Development, urging him to consider how to apply his office's colossal energies to peacetime. "New frontiers of the mind are before us," wrote Roosevelt. "And if they are pioneered with the same vision, boldness, and drive with which we have waged this war, we can create a fuller and more fruitful employment and a fuller and more fruitful life."[4]

The worlds of 1944 and today seem eons apart, but there are important similarities. Then, as now, technology was transforming the way people lived and worked. Then, as now, some feared technology's potential to inflict harm, while others considered it humanity's salvation. Then, as now, government and higher education sought to nurture and harness the power of progress. And in 1944, they had a clear idea of how to do it: they would forge a social compact.

Dr. Bush's reply to FDR took the form of a now-famous report, entitled *Science: The Endless Frontier*. Laying out a plan for the future of university-based research, Bush observed that "[s]cientific progress is one essential key to our security as a nation, to our better health, to more jobs, to a higher standard of living, and to our cultural progress."[5] To achieve these outcomes, he proposed a system in which government funding would flow into universities for four key purposes: to create new knowledge, educate the next generation of scientists, create new products and industries, and advance the public welfare. As Bush observed, "The rewards of such exploration both for the Nation and the individual are great."

Bush's words proved correct. The social compact between government and higher education that was established in the 1940s ushered in a new era of prosperity in U.S. society. Government delivered funding, while universities delivered new knowledge that was transformed into new jobs and new industries. The result was the foundation for much of the economic and technological progress of the twentieth century.

As with the dawn of the atomic age, the threshold of the AI age also demands a social compact that harnesses knowledge and learning to the service of a broader good. But its overarching goals should align with today's economic and societal impera-tives. Among other things, it will need to combat society's growing inequality by preparing people for the opportunities and challenges offered by a world of miraculous advances. As technology forges ahead, it is extremely likely that inequality will worsen. People who own advanced machines and capital will benefit immensely, while displaced human employees will be at strong risk of losing their livelihoods. In a world in which today, a mere eight people own an equal amount of wealth to one half of the planet's population, technology seems poised to tip the scales even further.[6]

Just as the social compact's goals should expand with the times, so can the roster of players who participate in it. For today's compact to be effective in light of radical technological change, employers should be included as full participants.

THE ROLE OF EMPLOYERS

Today, the relationship between higher education and employers is all too often a loose coupling, lacking in connection. Typically, interactions between universities and employers are surface-level and episodic, with businesses perhaps sending a manager to campus a few times of year to sit on an advisory panel for a degree program or discipline. Colleges and universities attempt to ascer-tain the needs of business but sometimes misread the signals, leading to a skills gap between graduates and the job market. A much tighter coupling, in which there is no miscommunication, is needed.

Under a new social compact, higher education and employers have the opportunity to integrate their activities much further, with industry fully participating in designing programs and producing them. As in the case of General Electric working with Northeastern to develop our in-house program in advanced manufacturing, employers and universities can move from a place of isolation from each other to one in which their enterprises are embedded together. Since the professional workplace is one of the key settings in which the impact of technological change will be felt most strongly, educational efforts meant to address those changes—theoretical and experiential learning alike—can take place in companies and offices, not just within the ivory tower.

We can see another form of innovative collaboration in play at San Jose State University, which in 2012 partnered with IBM to develop a program in "social business," teaching students to parlay their social networking savvy into job-ready skills.[7] IBM provided expertise, mentorship, and technology to help educate future members of the workforce in an area of pressing business need. Or consider the example of Drake University in Iowa. Known mostly for its liberal arts programs, Drake is now moving to integrate its academics more closely with demands in the job market. Lately, it has partnered with local industries to construct a data analytics major and minor "with business needs in mind."[8] They asked employers what core competencies they sought in employees, then mapped those competencies to existing programs, thereby revealing the gaps. They also engaged corporations in sponsoring the start-up costs for the new program. This sort of open communication and give-and-take is crucial if businesses and universities are to unite in fulfilling the social compact.

GLOBAL LESSONS

Of course, there are myriad approaches to generating social and economic progress by fostering collaboration between higher education and employers. Many other countries are pursuing innovative ways to deepen that collaboration, sometimes starting from a place of long-established integration.

Austria, Germany, and Switzerland, for example, remain the torchbearers of apprenticeship, and they continue to build on, and update, their educational systems for professional workplace education. In Austria, teachers in professionally oriented programs are required to have industry experience, and many juggle positions in teaching and in industry. Germany upholds its robust "dual system" of education by strongly engaging employers and social partners, as well as ensuring that it serves broader social and economic goals instead of short-term hiring needs. To accomplish this, they also prioritize research that focuses on the link between education and employment, funding a national network of research centers to study and improve their career-training system. For its part, Switzerland has prioritized making the value proposition of its education-employment collaborations robust, by keeping the costs to employers to participate in such programs much lower than the benefits they receive from apprentices and educational institutions.[9]

Ireland, which has a less comprehensive tradition of apprenticeship than other countries, is attempting to make up ground through a program dubbed the "National Skills Strategy 2025." One of the cornerstones of this plan is to increase employer engagement with education, creating 50,000 new apprenticeship and traineeship positions by 2020. It also pledges to more than double national participation in lifelong learning.[10] Of course, paying for this sort of ambitious initiative can be problematic. Hungary has answered its own need for increased career-oriented education through a national levy, which yields a stable source of funding for employer-focused education, funneling much of the revenue into experiential learning opportunities within business enterprises.[11]

A unique situation is currently unfolding in China: far from fearing the robot revolution, national leaders are embracing it. Investing billions of renminbi in robotics, Chinese leaders have launched an active program of mass automation across the country's industrial sector. Over the next few years, it is poised to integrate thousands of advanced robots into factories currently staffed mostly by human workers. Guangdong, the country's foremost manufacturing center, aims to automate 80 percent of its factories by 2020.[12] The goal of this vast investment in technology is to shift the entire Chinese economy from its manufacturing base to a service one—not dissimilar to the way current knowledge economies around the world are shifting to an AI economy. Instead of remaining the world's factory, China now aspires to become the world's bank, IT department, and human-resources desk.

To effectuate this transition successfully, Chinese governmental leaders and businesses will need to join with universities in preparing workers for their future roles. They might find a useful model in South Korea, where industry works hand in glove with universities in both education and research. There, it is common for industry to sponsor university programs, and regional universities receive earmarked financial grants to facilitate collaboration with employers. Moreover, South Korea boasts the highest proportion of joint university-industry research in the world—seven of the top fourteen collaborations with individual companies are between South Korean universities and the national conglomerate Samsung; a full 23 percent of published research from Pohang University of Science and Technology comes from industry collaborations.[13]

Clearly, the idea that universities and employers should work together for the common good is a global one. And while the specifics of these compacts and arrangements vary between nations, the goal of social and economic progress is universal. As the United States moves toward redefining its own country's interrelationships among higher education, industry, and government, it will no doubt be instructive to keep an eye on how other countries are responding to the challenges and opportunities of the AI age.

FALSE CHOICES, AND A REAL ONE

Much of the popular discourse about colleges and universities has, for good or ill, centered on the question of what sort of education is best—a discussion that is often reduced to the dichotomy between learning to live versus learning to earn a living, or between the value of a liberal arts education versus the value of "practical" course of study that promotes employability. However, the robot-proof education model shows that these are false choices.

The arrival of brilliant machines conclusively dispels the notion that a remunerative career is predicated on the study of an applied, "practical" subject. Going forward, machines will perform much of the work that was once the concern of these subjects, such as the simple analysis and application of facts to situations, or the management of data. Instead, the jobs of the future will demand the higher-order cognitive abilities and skills that are often associated with a liberal arts education, and that are pointedly inculcated through an education in humanics. As discussed throughout this book, the roles that human beings fill will be largely concerned with creativity.

It is also time to move beyond the canard that students must choose between an economically rewarding career and a fulfilling, elevated inner life. More than ever before, the capacities that equip people to succeed professionally are the same as the virtues espoused by Cardinal Newman in his paeans to "liberal knowledge"—namely, an agile mind, refinement of thought, and facility of expression.[14] By appropriating our routine work, machines will offer human beings a chance at liberation from drudgery, freeing us for more creative employment. Tomorrow's jobs will require us to deploy our creativity and higher-order capacities in the service of our companies, our economy, and society. Thus, in the past, a factory hand would have spent his days in rote manufacturing; tomorrow, he may have the opportunity to work as a skilled artisan within a company. Whereas a middle manager once engaged in pushing information between boxes, tomorrow she'll be an in-house entrepreneur. Likewise, the gray "Organization Man" described by William Whyte in the 1950s used to punch the clock and unthinkingly enforce company policies.[15] Tomorrow, he'll be developing his company's global business strategies and figuring out how to deploy the company's automated workforce to do it.

Whether the objective is to educate younger learners, older learners, company employees, or gig workers, the bottom line is the same. Learning is now an ongoing voyage with many ports of call but never a terminus. This continuous journey has implications that reach far past the campus gates, into our homes, our workplaces, and our startups. These implications will shape our ambitions and even our laws. Ultimately, they will affect everyone, carried throughout the world by multi-university networks built to transcend limitations of scale, place, and time.

Education is not a panacea for humanity's troubles. We cannot educate ourselves out of all our social and natural predicaments. We can; however, help individuals brace for change and embrace the technological miracles that lie ahead. Perhaps, if we educate enough of them, society's weight will shift, making it more equitable, more just, and more sustainable. I believe that when people are given education, they may still be astonished by the changes and mysteries that the future holds, but they will see these as opportunities rather than threats.

Such a world, I believe, is possible. It is our job to make it happen.

NOTES

Introduction

1. Eric Brynjolfsson and Andrew McAfee, *The Second Machine Age: Work, Progress, and Prosperity in a Time of Brilliant Technologies* (New York: Norton, 2014), 9.

2. Martin Ford, *Rise of the Robots: Technology and the Threat of a Jobless Future* (New York: Basic Books, 2015), xvi.

3. Alistair Gray, "MasterCard to Start Trialing Pepper the Robot in Pizza Hut," *Financial Times*, May 24, 2016, https://www.ft.com/content/2b78d806-20f2-11e6-aa98-db1e01fabc0c.

4. Geoff Dyer, "US to Deploy Robot Combat Strategists," *Financial Times*, April 27, 2016, https://www.ft.com/content/29b93562-0c5f-11e6-b0f1-61f222853ff3.

5. Michael Pooler, "Industrial Robot Sales Hit Record," *Financial Times*, June 22, 2016, https://www.ft.com/content/d8d80f32-3874-11e6-a780-b48ed7b6126f.

6. Ravi Kalakota, "Love, Sex and Predictive Analytics: Tinder, Match.com, and OkCupid," *Business Analytics 3.0* (blog), May 29, 2015, https://practicalanalytics.co/2015/05/29/love-sex-and-predictive-analytics-tinder-match-com-and-okcupid.

7. Elizabeth Lopatto, "Google's AI Can Learn to Play Video Games," *The Verge*, February 25, 2015, http://www.theverge.com/2015/2/25/8108399/ google-ai-deepmind-video-games.

8. Stanley C. Litow, personal interview, February 24, 2016.

9. Lopatto, "Google's AI Can Learn to Play Video Games."

10. Carl Benedikt Frey and Michael Osborne, "The Future of Employment: How Susceptible Are Jobs to Computerisation?," Oxford Martin School, University of Oxford, September, 2013, http://www.oxfordmartin.ox.ac.uk/ downloads/academic/The_Future_of_Employment.pdf.

11. Nathaniel Popper, "The Robots Are Coming for Wall Street," *New York Times Magazine*, February 25, 2016, http://www.nytimes.com/2016/02/28/ magazine/the-robots-are-coming-for-wall-street.html?smprod=nytcore -ipad&smid=nytcore-ipad-share.

12. Michael Chui, James Manyika, and Medhi Miremadi, "Four Fundamentals of Workplace Automation," *McKinsey Quarterly*, November, 2015, http://www.mckinsey.com/business-functions/business-technology/ our-insights/four-fundamentals-of-workplace-automation.

13. Ryan Avent, *The Wealth of Humans: Work, Power, and Status in the Twenty-first Century* (New York: St. Martin's Press, 2016), 5.

14. U.S. Bureau of Labor Statistics, *Occupational Outlook Handbook*, http:// www.bls.gov/ooh/computer-and-information-technology/home.htm.

15. Danny Hakim, "GM Will Reduce Hourly Workers in US by 25,000," *New York Times*, June 8, 2005, http://www.nytimes.com/2005/06/08/ business/gm-will-reduce-hourly-workers-in-us-by-25000.html.

16. Eric Morath, "Gig Economy Attracts Many Workers, Few Full-Time Jobs," *Wall Street Journal*, February 18, 2016, http://blogs.wsj.com/economics/ 2016/02/18/gig-economy-attracts-many-workers-few-full-time-jobs.

17. General Assembly & Burning Glass Technologies, "Hybrid Jobs: Blurring Lines—How Business and Technology Skills Are Merging to Create

High Opportunity Hybrid Jobs," 2015, http://burning-glass.com/research/hybrid-jobs.

18. United Nations Department of Economic and Social Affairs, *World Population Prospects, the 2015 Revision*, July 29, 2015, https://esa.un.org/unpd/wpp.

Chapter 1

1. "The Chapman University Survey of American Fears," 2016, http://www.chapman.edu/wilkinson/research-centers/babbie-center/survey-american-fears.aspx.

2. Kenneth Miller, "Archaeologists Find Earliest Evidence of Humans Cooking with Fire," *Discover Magazine*, December 17, 2013, http://discovermagazine.com/2013/may/09-archaeologists-find-earliest-evidence-of-humans-cooking-with-fire.

3. Yuval Noah Harari, *Sapiens: A Brief History of Humankind* (New York: HarperCollins, 2015), 12.

4. Eric Hobsbawm, "The Machine Breakers," *Past and Present* 1(1) (1952): 57–70, http://web.csulb.edu/~ssayeghc/theory/wintertheory/machinebreakers.pdf.

5. Lord Byron's Speech, *Luddites at 200*, http://www.luddites200.org.uk/LordByronspeech.html.

6. "An Ode to the Framers of the Frame Bill," *Luddites at 200*, http://www.luddites200.org.uk/documents/Byronpoems.pdf.

7. Erik Brynjolfsson and Andrew McAfee, "Will Humans Go the Way of Horses?," *Foreign Affairs*, July–August 2015, 8.

8. Ibid.

9. John Henry Newman, "Discourse 7. Knowledge Viewed in Relation to Professional Skill," in *The Idea of a University*, 1858, 178, http://www.newmanreader.org/works/idea/discourse7.html.

10. The 1890 Land Grant Universities, "The Morrill Acts of 1862 and 1890," 2015, http://www.1890universities.org/history.

11. Clark Kerr, *The Uses of the University*, 5th ed. (Cambridge, MA: Harvard University Press, 2001), 3.

12. Suzanne Mettler, "How the G.I. Bill Built the Middle Class and Enhanced Democracy," *The Scholars Strategy Network*, January 2012, http://www.scholarsstrategynetwork.org/brief/how-gi-bill-built-middle-class-and-enhanced-democracy.

13. Eliza Berman, "How the G.I. Bill Changed the Face of Higher Education in America," *Time*, June 22, 2015, http://time.com/3915231/student-veterans.

14. Kerr, *The Uses of the University*, 36.

15. American Association for the Advancement of Science, "R&D at Colleges and Universities," updated March 24, 2016, https://www.aaas.org/page/rd-colleges-and-universities.

16. United States Census Bureau, "World Population: Historical Estimates of World Population," updated December, 2013, http://www.census.gov/population/international/data/worldpop/table_history.php.

17. Population Reference Bureau, "Human Population: Urbanization," 2016, http://www.prb.org/Publications/Lesson-Plans/HumanPopulation/Urbanization.aspx.

18. Jonathan James, "The College Wage Premium," Federal Reserve Bank of Cleveland, August 8, 2012, https://www.clevelandfed.org/newsroom-and-events/publications/economic-commentary/2012-economic-commentaries/ec-201210-the-college-wage-premium.aspx.

19. Nathaniel Popper, "The Robots Are Coming for Wall Street," *New York Times Magazine*, February 25, 2016, http://www.nytimes.com/2016/02/28/magazine/the-robots-are-coming-for-wall-street.html?smprod=nytcore-ipad&smid=nytcore-ipad-share.

20. Clive Cookson, "US Researchers Enter the Cutting Edge with First Robot Surgeon," *Financial Times*, May 4, 2016, https://www.ft.com/content/d23c7a4e-11d9-11e6-91da-096d89bd2173.

21. Scott Semel, Intralinks, personal interview, March 14, 2016.

22. Klaus Schwab, *The Fourth Industrial Revolution* (Geneva: World Economic Forum, 2016).

23. Colin Angle, iRobot Corporation, personal interview, April 4, 2016.

24. Martin Ford, *Rise of the Robots: Technology and the Threat of a Jobless Future* (New York: Basic Books, 2015), 75.

25. Ryan Avent, *The Wealth of Humans: Work, Power, and Status in the Twenty-first Century* (New York: St. Martin's Press, 2016), 5.

26. Richard Henderson, "Industry Employment and Output Projections to 2020," *Monthly Labor Review*, January 2012, 66, http://www.bls.gov/opub/mlr/2012/01/art4full.pdf.

27. Richard Henderson, "Industry Employment and Output Projections to 2024," *Monthly Labor Review*, December 2015, 2, http://www.bls.gov/emp/ep_table_201.htm.

28. Facebook, "Stats," http://newsroom.fb.com/company-info.

29. Twitter, "What's Happening," https://about.twitter.com/company.

30. Nir Jaimovich and Henry E. Siu, "The Trend Is the Cycle: Job Polarization and Jobless Recoveries," Working Paper No. 18334, National Bureau of Economic Research, August 2012, http://www.nber.org/papers/w18334.

31. PwC, "Work-Life 3.0: Understanding How We'll Work Next," PwC Consumer Intelligence Series, 2016, 11, https://www.pwc.com/us/en/industry/entertainment-media/publications/consumer-intelligence-series/assets/pwc-consumer-intellgience-series-future-of-work-june-2016.pdf.

32. Emily Symkal, "Flexible Jobs: How the Gig Economy Is Impacting Recruiting," March 7, 2016, https://www.jibe.com/blog/what-recruiters-should-know-about-flexible-jobs-today.

33. Cecilia Kang, "No Driver? Bring It On. How Pittsburgh Became Uber's Testing Ground," *New York Times*, September 10, 2016, http://www.nytimes.com/2016/09/11/technology/no-driver-bring-it-on-how-pittsburgh-became-ubers-testing-ground.html.

34. Josh Bivens, Elise Gould, and Lawrence Mishel, "Wage Stagnation in Nine Charts," Economic Policy Institute, January 6, 2015, http://www.epi.org/publication/charting-wage-stagnation.

35. National Center for Education Statistics, "Postbaccalaureate Enrollment," *The Condition of Education*, May 2016, http://nces.ed.gov/programs/coe/indicator_chb.asp.

36. Karen Turner, "Why Students Are Throwing Tons of Money at a Program That Won't Give Them a College Degree," *Washington Post*, March 17, 2016, https://www.washingtonpost.com/news/the-switch/wp/2016/03/17/why-students-are-throwing-tons-of-money-at-a-program-that-wont-give-them-a-college-degree.

37. Jason Tanz, "Soon We Won't Program Computers. We'll Train Them Like Dogs," *Wired*, May 17, 2016, https://www.wired.com/2016/05/the-end-of-code.

38. Harari, *Sapiens*, 24.

39. Ibid., 25.

40. Ibid.

Chapter 2

1. Charles Duhigg and Keith Bradsher, "How the U.S. Lost Out on iPhone Work," *New York Times*, January 21, 2012, http://www.nytimes.com/2012/01/22/business/apple-america-and-a-squeezed-middle-class.html?_r=1&hp=&pagewanted=all.

2. Simon Parry, "The True Cost of Your Cheap Clothes: Slave Wages for Bangladesh Factory Workers," *South China Morning Post Magazine*, June 11, 2016, http://www.scmp.com/magazines/post-magazine/article/1970431/ true-cost-your-cheap-clothes-slave-wages-bangladesh-factory.

3. Diana Smeltz, Craig Kafura, and Lily Wojtowicz, "Actually, Americans Like Free Trade," Chicago Council on Global Affairs, September 7, 2016, https://www.thechicagocouncil.org/publication/actually-americans-free -trade.

4. Pew Research Center, "The American Middle Class Is Losing Ground," December 9, 2015, 4, http://www.pewsocialtrends.org/files/2015/12/2015- 12-09_middle-class_FINAL-report.pdf.

5. Emmanuel Saez, "U.S. Top One Percent of Income Earners Hit New High in 2015 amid Strong Economic Growth," Washington Center for Equitable Growth, July 1, 2016, http://equitablegrowth.org/research-analysis/ u-s-top-one-percent-of-income-earners-hit-new-high-in-2015-amid-strong -economic-growth.

6. Pew Research Center, "The American Middle Class Is Losing Ground," 1.

7. Ibid., 2.

8. Bureau of Labor Statistics, table 1.4, Occupations with the Most Job Growth, April 18, 2016, http://www.bls.gov/emp/ep_table_104.htm.

9. Pew Research Center, "The State of American Jobs," October 6, 2016, http://www.pewsocialtrends.org/2016/10/06/the-state-of-american-jobs.

10. National Association of Colleges and Employers, "Job Outlook 2016: Attributes Employers Want to See on New College Graduates' Resumes," November 18, 2015, http://www.naceweb.org/s11182015/employers-look-for -in-new-hires.aspx.

11. Northeastern University and FTI Consulting, Business Elite National Poll, Third Installment of the Innovation Imperative Polling Series, Topline

Report, survey conducted February 3–19, 2014, http://www.northeastern
.edu/innovationsurvey/pdfs/Pipeline_toplines.pdf.

12. James Bessen, "Employers Aren't Just Whining: The 'Skills Gap' Is
Real," *Harvard Business Review*, August 25, 2014, https://hbr.org/2014/08/
employers-arent-just-whining-the-skills-gap-is-real.

13. Executive Office of the President, National Science and Technology
Council Committee on Technology, "Preparing for the Future of Artificial
Intelligence," October 2016, 5.

14. Ibid., 8.

15. David Julian, Wells Fargo, personal interview, March 3, 2016.

16. Nathaniel Popper, "The Robots Are Coming for Wall Street," *New York
Times Magazine*, February 25, 2016, http://www.nytimes.com/2016/02/28/
magazine/the-robots-are-coming-for-wall-street.html?smprod=nytcore
-ipad&smid=nytcore-ipad-share.

17. See Richard Susskind and Daniel Susskind, *The Future of the Professions:
How Technology Will Transform the Work of Human Experts* (Oxford, UK:
Oxford University Press, 2015), for a good treatment of how professions
such as law, medicine, and others may be affected further by oncoming
technological change.

18. Grant Theron, Young & Rubicam, personal interview, February 23,
2016.

19. William Manfredi, Young & Rubicam, personal interview, February
23, 2016.

20. Peter McCabe, GE Transportation, personal interview, February 11,
2016.

21. Andrea Cox, GE Aviation, personal interview, February 23, 2016.

22. Steve Vinter, Google, personal interview, March 9, 2016.

23. Darren Donovan, KPMG, personal interview, March 2, 2016.

24. Sjoerd Gehring, Johnson & Johnson, personal interview, March 3, 2016.

25. Marc Andreesen, "Why Software Is Eating the World," *Wall Street Journal*, August 20, 2011.

Chapter 3

1. Catherine Fisher, "LinkedIn Unveils the Top Skills That Can Get You Hired In 2017, Offers Free Courses for a Week," *LinkedIn Official Blog*, October 20, 2016, https://blog.linkedin.com/2016/10/20/top-skills-2016 -week-of-learning-linkedin.

2. Martin Ford, *Rise of the Robots: Technology and the Threat of a Jobless Future* (New York: Basic Books, 2015), 256.

3. Ryan Avent, *The Wealth of Humans: Work, Power, and Status in the Twenty-first Century* (New York: St. Martin's Press, 2016), 64.

4. Ibid., 5.

5. Carl Haub, "How Many People Have Ever Lived on Earth?," Population Reference Bureau, October 2011, http://www.prb.org/Publications/Articles/ 2002/HowManyPeopleHaveEverLivedonEarth.aspx.

6. Amy X. Wang, "The Musical AI Is Now Working on Its Debut Album(s)— and Wants to Do the Beatles Better Than the Beatles," *Quartz*, October 18, 2016, http://qz.com/812231/sony-is-making-an-artificial-intelligence-algorithm -that-writes-perfect-hit-making-songs.

7. Kyung Hee Kim, "Can We Trust Creativity Tests? A Review of the Torrance Tests of Creative Thinking (TTCT)," *Creativity Research Journal* 18(1) (2006): 3, http://people.uncw.edu/caropresoe/GiftedFoundations/ SocialEmotional/Creativity-articles/Kim_Can-we-trust-creativity-tests.pdf.

8. J. P. Guilford, *The Nature of Human Intelligence* (New York: McGraw-Hill, 1967).

9. Kim, "Can We Trust Creativity Tests?," 4.

10. Ford, *Rise of the Robots*, 130.

11. Ken Robinson, "Do Schools Kill Creativity?," Talk delivered at TED 2006, Conference on The Future We Will Create, Monterrey, California, February 2006, https://www.ted.com/playlists/171/the_most_popular_talks _of_all.

12. Richard Arum and Josipa Roksa, *Academically Adrift: Limited Learning on College Campuses* (Chicago: University of Chicago Press, 2011), 121.

13. National Center for Education Statistics, "Skills of U.S. Unemployed, Young, and Older Adults in Sharper Focus: Results from the Program for the International Assessment of Adult Competencies (PIAAC) 2012/2014," March 2016, http://nces.ed.gov/pubs2016/2016039.pdf.

14. The discipline of humanics is articulated in "Northeastern 2025," Northeastern University's 2016 academic plan (http://www.northeastern .edu/academic-plan).

15. In discussing the idea of a set of new literacies, it is necessary to acknowledge the work of others who have previously articulated similar concepts. In particular, the work of economists Frank Levy and Richard Murnane— who a generation ago articulated the concept of the new basic skills needed for students to meet the challenges of the knowledge economy—is key. So too are the contributions of media scholar Henry Jenkins, who has defined the concept of new media literacies to enable people to navigate the complexities of the digital media age. See, e.g., Frank Levy and Richard J. Murnane, R., *The New Division of Labor: How Computers Are Creating the Next Job Market* (Princeton, NJ: Princeton University Press, 2004), and Henry Jenkins, *Convergence Culture: Where Old and New Media Collide* (New York: New York University Press, 2006).

16. Liz Eggleston, "Coding Bootcamp Market Sizing Report 2016," *Course Report*, June 22, 2016, https://www.coursereport.com/reports/2016-coding -bootcamp-market-size-research.

17. Sjoerd Gehring, Johnson & Johnson personal interview, March 3, 2016.

18. David S. Bennahum, "Coding Snobs Are Not Helping Our Children Prepare for the Future," *Quartz*, June 10, 2016, http://qz.com/703335/coding-snobs-are-not-helping-our-children-prepare-for-the-future.

19. Dave Evans, "The Internet of Things: How the Next Evolution of the Internet Is Changing Everything," Cisco Internet Business Solutions Group, April 2011, http://www.cisco.com/c/dam/en_us/about/ac79/docs/innov/IoT_IBSG_0411FINAL.pdf.

20. "Caterpillar and the Internet of Big Things," Caterpillar, October 15, 2015, http://www.caterpillar.com/en/news/caterpillarNews/innovation/caterpillar-disrupted.html.

21. Michael Patrick Lynch, *The Internet of Us: Knowing More and Understanding Less in the Age of Big Data* (New York: Liveright, 2016), 161.

22. Ibid., 164.

23. Geoff Colvin, *Humans Are Underrated: What High Achievers Know That Brilliant Machines Never Will* (New York: Portfolio/Penguin, 2015), 121.

24. Ibid., 128.

25. Marcia S. Smith, "NASA's Space Shuttle *Columbia*: Synopsis of the Report of the *Columbia* Accident Investigation Board," CRS Report for Congress, Congressional Research Service, Library of Congress, September 2, 2003, https://history.nasa.gov/columbia/Troxell/Columbia%20Web%20Site/Documents/Congress/CRS%20Summary%20of%20CAIB%20Report.pdf.

26. Christopher F. Schuetze, "A Dutch Architect Offshores the Future of Housing," *New York Times*, November 28, 2016, http://www.nytimes.com/2016/11/28/arts/design/offshoring-the-future-of-housing.html?_r=0.

27. Waters Foundation, http://watersfoundation.org.

28. World Economic Forum, "Human Capital Outlook: Association of Southeast Asian Nations (ASEAN)," WEF, Kuala Lumpur, Malaysia, June 1–2, 2016, http://www3.weforum.org/docs/WEF_ASEAN _HumanCapitalOutlook.pdf.

29. U.S. Bureau of Labor Statistics, "Entrepreneurship and the U.S. Economy," 2015, http://www.bls.gov/bdm/entrepreneurship/entrepreneurship.htm.

30. Desh Deshpande, Sycamore Networks and Deshpande Foundation, personal interview, March 2, 2016.

31. FailCon, http://thefailcon.com/about.html.

32. Paula Caligiuri, *Cultural Agility: Building a Pipeline of Successful Global Professionals* (San Francisco: Jossey-Bass, 2012), 4.

33. J. Manyika, S. Lund, J. Bughin, J. Woetzel, K. Stamenov, and D. Dhingra, "Digital Globalization: The New Era of Global Flows," McKinsey Global Institute, February 2016, http://www.mckinsey.com/business -functions/mckinsey-digital/our-insights/digital-globalization-the-new -era-of-global-flows.

34. Ibid., 4.

35. Ibid., 6.

36. Diksha Madhok, "The Story behind India's Rice Bucket Challenge," *Quartz*, August 25, 2014, http://qz.com/254910/india-adapts-the-ice -bucket-challenge-to-suit-local-conditions-meet-the-rice-bucket-challenge.

37. Jie Zong and Jeanne Batalova, "Frequently Requested Statistics on Immigrants and Immigration in the United States," Migration Policy Institute, April 14, 2016, http://www.migrationpolicy.org/article/frequently -requested-statistics-immigrants-and-immigration-united-states.

38. Caligiuri, *Cultural Agility*, 6.

39. Raffi Khatchadourian, "We Know How You Feel," *New Yorker*, January 19, 2015, http://www.newyorker.com/magazine/2015/01/19/know-feel.

Chapter 4

1. Dawn Kawamoto, "Watson Wasn't Perfect: IBM Explains the *Jeopardy!* Errors," *Aol.com*, February 17, 2011, http://www.aol.com/article/2011/02/17/ the-watson-supercomputer-isnt-always-perfect-you-say-tomato/19848213.

2. Ken Jennings, "My Puny Human Brain," *Slate*, February 16, 2011, http:// www.slate.com/articles/arts/culturebox/2011/02/my_puny_human_brain .html.

3. John Dewey, *Experience and Education* (New York: Touchstone, 1938), http://ruby.fgcu.edu/courses/ndemers/colloquium/experienceducationdewey .pdf.

4. Ibid.

5. Ronald Fry and David Kolb, "Experiential Learning Theory and Learning Experiences in Liberal Arts Education," *New Directions for Experiential Learning*, No. 6, *Enriching the Liberal Arts through Experiential Learning* (San Francisco: Jossey-Bass, 1979), 80.

6. The contributions of my colleague Susan Ambrose, Northeastern's senior vice provost for undergraduate education and experiential learning, to the field of learning science have been critical to the ideas developed in this section. I thank her for her assistance.

7. Susan A. Ambrose, Michael W. Bridges, Michele DiPietro, Marsha C. Lovett, Marie K. Norman, and Richard E. Mayer, *How Learning Works: Seven Research-Based Principles for Smart Teaching* (San Francisco: Jossey-Bass, 2010), 95.

8. Ibid., 97.

9. Ibid., 108.

10. Ibid., 110.

11. Ibid., 110.

12. Carol S. Dweck, *Mindset: The New Psychology of Success* (New York: Ballantine Books, 2008), 12.

13. Northeastern University, "All College Grads Want to Be Prepared for Their Careers. Northeastern Students Actually Are," http://www .northeastern.edu/preparedness.

14. Northeastern University and FTI Consulting, Business Elite National Poll, Third Installment of the Innovative Imperative Polling Series, Topline Report, survey conducted February 3–19, 2014, http://www.northeastern .edu/innovationsurvey/pdfs/Pipeline_toplines.pdf.

15. Northeastern University, Innovation in Higher Education Survey Toplines, survey conducted October 13–18, 2012, http://www.northeastern .edu/test/innovationsurvey/pdfs/survey-results.pdf.

16. Catherine Erdelyi, personal interview, February 9, 2016.

17. McKenzie Jones, personal interview, February 11, 2016.

18. Mary Tobin, personal interview, February 22, 2016.

19. Ali Matalon, personal interview, February 5, 2016.

20. Ramanda Nanda and Jesper B. Sorenson, "Workplace Peers and Entre-preneurship," *Management Science* 56(7) (2010): 1116, http://www.edegan .com/pdfs/Nanda%20Sorensen%20(2010)%20-%20Workplace%20 peers%20and%20entrepreneurship.pdf.

21. Edward P. Lazear, "Entrepreneurship," *Journal of Labor Economics* 23(4) (2005): 649–680, http://www2.econ.iastate.edu/classes/econ521/orazem/ Papers/Lazear_entrepreneurship.pdf.

22. Portions of this section appeared, in part, in "A Complete Education," *Inside Higher Ed*, April 20, 2015.

23. Uta Poiger, dean of Northeastern's College of Social Sciences and Humanities, has been instrumental in articulating the concept of "experiential liberal arts" at our university. The discussion in this section draws deeply on her insights.

24. Noah E. Friedkin, Anton V. Proskurnikov, Roberto Tempo, and Sergey E. Parsegov, "Network Science on Belief System Dynamics under Logic Constraints," *Science*, October 21, 2016, 321–326, science.sciencemag.org/content/354/6310/321.

25. "Purposeful Work," Bates College website, https://www.bates.edu/purposeful-work.

26. I again acknowledge Susan Ambrose and her colleagues in Northeastern's Center for the Advancement of Teaching and Learning Through Research, who developed the SAIL app.

Chapter 5

1. Arnold J. Toynbee, *A Study of History*, vol. 1, *Introduction: The Geneses of Civilizations* (London: Oxford University Press, 1934), 24.

2. Jared Diamond, *Collapse: How Societies Choose to Fail or Succeed* (New York: Viking Press, 2005), 274–275.

3. Alan Tait, "Reflections on Student Support in Open and Distance Learning." *International Review of Research in Open and Distance Learning* (2003), http://oro.open.ac.uk/1017/1/604.pdf.

4. Daniel Webster, "Lecture before the Society for the Diffusion of Useful Knowledge, Boston, November 11, 1836," *The Writings and Speeches of Daniel Webster*, vol. 13 (Boston: Little, Brown, 1903), Google Books, https://books.google.com/books?id=2iF3AAAAMAAJ&pg=PA63&source=gbs_toc_r&cad=3#v=onepage&q&f=false; Oliver Wendell Holmes, *Currents and Counter-Currents in Medical Science: With Other Addresses and Essays* (Boston: Ticknor and Fields, 1861), Google Books, https://books.google.com/books?id=c8MNAAAAYAAJ&printsec=frontcover&source=gbs_ge_summary_r&cad=0#v=onepage&q&f=false.

5. "About the Lowell Institute," www.lowellinstitute.org/about.

6. YMCA, "History—Founding," http://www.ymca.net/history/founding.html.

7. Institute of Education Sciences (IES) and National Center for Educational Statistics (NCES), "Fast Facts: Back to School Statistics," https://nces .ed.gov/fastfacts/display.asp?id=372.

8. Institute of Education Sciences (IES) and National Center for Educational Statistics (NCES), "Table 303.40. Total fall enrollment in degree-granting postsecondary institutions, by attendance status, sex, and age: Selected years 1970 through 2025," 2015, https://nces.ed.gov/programs/ digest/d15/tables/dt15_303.40.asp?current=yes.

9. American Association of Community Colleges, "2014 Fact Sheet," http://www.aacc.nche.edu/AboutCC/Documents/Facts14_Data_R3.pdf.

10. Institute of Education Sciences (IES) and National Center for Educational Statistics (NCES), "Web Tables: U.S. Department of Education," December 2011, NCES 2012-173, https://nces.ed.gov/pubs2012/2012173.pdf.

11. Patrick Gillespie, "University of Phoenix Has Lost Half Its Students," *CNNMoney*, March 15, 2015, http://money.cnn.com/2015/03/25/investing/ university-of-phoenix-apollo-earnings-tank.

12. "Keeping It on the Company Campus," *The Economist*, May 16, 2015, http://www.economist.com/news/business/21651217-more-firms-have-set -up-their-own-corporate-universities-they-have-become-less-willing-pay.

13. Philipp Kolo, Ranier Strack, Philippe Cavat, Roselinde Torres, and Vikram Bhalia, "Corporate Universities: An Engine for Human Capital," *BCG Perspectives*, July 18, 2013, https://www.bcgperspectives.com/content/ articles/human_resources_leadership_talent_corporate_universities _engine_human_capital/#chapter1.

14. "Cognition Switch: What Employers Can Do to Encourage Their Workers to Retrain," Special Report on Lifelong Education, *The Economist*, January 14, 2017, 9.

15. Isabel Carndenas-Navia and Brian Fitzgerald, "The Broad Application of Data Science and Analytics: Essential Tools for the Liberal Arts Graduate," *Change: The Magazine of Higher Learning*, July 31, 2015, http://www .tandfonline.com/doi/full/10.1080/00091383.2015.1053754.

16. Ryan Denham, "Illinois State Meets Growing Need for Cybersecurity Professionals," *Illinois State University News*, April 21, 2016, https://news.illinoisstate.edu/2016/04/illinois-state-meets-growing-need-cybersecurity-professionals.

17. "Ohio State among Select Schools to Launch Curriculum with IBM Watson," The Ohio State University College of Engineering, May 7, 2014, https://engineering.osu.edu/news/2014/05/ohio-state-among-select-schools-chosen-launch-curriculum-ibm-watson.

18. Barbara Means, Yuki Toyama, Robert Murphy, and Marianne Baki, "The Effectiveness of Online and Blended Learning: A Meta-Analysis of the Empirical Literature," *Teachers College Record* 115, March 2013, 2, https://www.sri.com/sites/default/files/publications/effectiveness_of_online_and_blended_learning.pdf.

19. Ibid.

20. Ian Hathaway and Mark Muro, "Tracking the Gig Economy: New Numbers," Brookings Institution, October 13, 2016, https://www.brookings.edu/research/tracking-the-gig-economy-new-numbers.

21. Nicholas Wells, "The 'Gig Eeconomy' Is Growing—and Now We Know by How Much," CNBC, October 13, 2016, http://www.cnbc.com/2016/10/13/gig-economy-is-growing-heres-how-much.html.

22. Laura G. Knapp, Janice E. Kelly-Reid, and Scott A. Ginder, "Employees in Postsecondary Institutions, Fall 2010, and Salaries of Full-time Instructional Staff, 2010–11," National Center for Educational Statistics, November 2011, p. 5, https://nces.ed.gov/pubs2012/2012276.pdf.

23. Ibid.

24. National Association of College and University Business Officers, "New Report Shows Recent Trends in Faculty Employment and Salaries," November 29, 2011, http://www.nacubo.org/Research/Research_News/New_Report_Shows_Recent_Trends_in_Faculty_Employment_and_Salaries.html.

25. Ibid.

26. Clark Kerr, *The Uses of the University* (Cambridge, MA: Harvard University Press, 1963).

27. Martin Ihrig and Ian MacMillian, "How to Get Ecosystem Buy-In," *Harvard Business Review*, March–April 2017.

Afterword

1. Economic Research, Federal Reserve Bank of St. Louis, "Unemployment Rate for United States," https://fred.stlouisfed.org/series/M0892BUSM156SNBR.

2. Centers for Disease Control and Prevention, "Leading Causes of Death, 1900–1998," https://www.cdc.gov/nchs/data/dvs/lead1900_98.pdf.

3. NOAA National Centers for Environmental Information, Climate at a Glance: Global Time Series, March 2017,http://www.ncdc.noaa.gov/cag.

4. Vannevar Bush, *Science: The Endless Frontier* (Washington, DC: United States Government Printing Office, 1945),https://nsf.gov/about/history/vbush1945.htm.

5. Ibid.

6. Oxfam International, "Just 8 Men Own Same Wealth as Half the World," January 17, 2017, https://www.oxfam.org/en/pressroom/pressreleases/2017-01-16/just-8-men-own-same-wealth-half-world.

7. "IBM and San Jose State University Collaborate to Advance Social Business Skills," IBM News Room, January 11, 2012, https://www-03.ibm.com/press/us/en/pressrelease/36486.wss.

8. "Equipping Liberal Arts Students with Skills in Data Analytics." Business-Higher Education Forum, 2016, http://www.bhef.com/sites/default/files/BHEF_2016_DSA_Liberal_Arts.pdf.

9. The information on Austria, Germany, and Switzerland is from OECD, "Learning for Jobs," OECD Reviews of Vocational Education

and Training, May 2011, https://www.oecd.org/edu/skills-beyond-school/
LearningForJobsPointersfor%20PolicyDevelopment.pdf.

10. Department of Education and Skills, "Government Launches Ireland's
National Skills Strategy 2025: Ireland's Future.", January 27, 2016, http://
www.education.ie/en/Press-Events/Press-Releases/2016-Press-Releases/
PR2016-01-27.html.

11. OECD, "Learning for Jobs."

12. John Pabon, and Lin Wang, "A New Era: Optimizing Chinese Industry in the Age of Automation," BSR, February 2017, https://www.bsr.org/
reports/BSR_Optimizing_Chinese_Industry_in_the_Age_of_Automation
.pdf.

13. John Morgan, "South Korean Universities Lead the Way on Industry Collaboration," *Times Higher Education* (March 9, 2017), https://www
.timeshighereducation.com/news/south-korean-universities-lead-way
-on-industry-collaboration.

14. Cardinal John Henry Newman, "The Idea of a University,178. http://
www.newmanreader.org/works/idea/discourse7.html.

15. William H. Whyte, Jr., *The Organization Man* (New York: Simon and
Schuster, 1956).

INDEX

Career services, importance of
alumni to, 133–134
Carnegie Mellon University,
global multi-university network
approach, 137
Change, resistance to, xvi–xvii,
1–3, 5–6, 8, 13–15, 46, 114
China, robotics in, x, 146
Classroom learning
and assessments/grading,
108–109
and content knowledge, 85,
89
and fostering growth mindsets,
89
integrating with real-world
experiences, xx, 75, 79–81, 86,
90, 93, 98
overemphasis on performance,
88–89
and teaching young children,
56, 66
Climate change, finding
solutions to, 43, 65–66,
139–140
Co-bots, x
Coding. *See* Technological
literacy
Cognitive capacities
approaches to teaching, 53–54,
73–75
critical thinking, 62
goal setting and evaluations,
108–109

in humanics curriculum,
overview, xix
Collaboration, as aspect of
human literacy, 59. *See
also* Colleges and
universities
Colleges and universities. *See
also* Experiential learning;
Humanics; Lifelong learning/
learners
alumni, 133–134
artificial divisions in curricula,
18, 129
and collaborations with
employers, 10, 122–124, 126,
141–147
and customized learning
programs, 121–127
debates about types of
education, 147
experiential liberal arts, 103–107,
148
faculty, 131–132
and federal funding for
research, 10–11
for-profit universities, 119
fund-raising, 134
and the G.I. Bill, 9–11
impact of Industrial Revolution,
8
individualized instruction
approaches, 112–113
lifelong learners vs. traditional
full-time students, 128

new literacies and cognitive capacities associated with, xviii–xiv, xix
systems thinking components, 64–66
and teaching where the learners are, xx
and technological literacy, 55–56
Human literacy
characteristics, xix, 58–61
and embracing diversity, 59–60
and handling ethical issues, 60–61
need for communication skills, 60
and working for economic equality/social justice, 61, 68
Humans. *See also* Creativity/creators
capacity for imagination, 64–65
and doing what machine cannot do, 19–20, 62–63, 65, 78–79, 87
and fears about robots/AI, 1–2, 5–6, 13–15, 46
and intelligence, 51–52
and liberating potential of machines, x, xvi–xvii, 2–3
Hungary, education-employment collaborations, 146
Hurricane Katrina, 64
Hybrid jobs, xv
Hybrid learning approaches, 125

IBM
Deep Blue supercomputer, 19
partnership in cognitive computing courses, 123
partnership with Memorial Sloan Kettering, xi
partnership with San Jose State University, 144
Watson supercomputer, xi, 77–78
Implants, transplants, unequal access to, 61
Income/economic inequality. *See* Social and economic justice; Wages
India, out-sourcing of legal work to, 31
Industrial Revolution
and displaced workers, 3, 5–5, 8, 12, 17, 26
and educational opportunity, 8, 114–115
and the expansion of human creativity, 14
and resistance to change, xvi–xvii, 1–3, 5–6, 8, 13–15, 46, 114
Information revolution, 4
Intelligent machines. *See also* Artificial intelligence (AI)/robotics; Automation/technology; Humans
ability to understand complex systems, 64–65

Intelligent machines (cont.)
deep learning, 90
and demands on educational
systems, 17
employment impacts; 14–15, 18,
41–43
ethical issues raised by, 60–61
liberating potential, x, xvi–xvii,
2–3
machine learning vs. human
learning, ix, 13, 78–80
and nonquantifiable thinking,
62–63
super-intelligent machines, 29
and uniquely human abilities
and skills, 19–20, 65, 78–79,
87
International vs. global
approaches, 138–139
Internet of Things, 15–16
*The Internet of Us: Knowing More
and Understanding Less in the
Age of Big Data* (Lynch), 58
Internships. *See also* Co-op
program, Northeastern
University; Experiential
learning; Lifelong learning/
learners
contrast with co-op program,
94
and experiential liberal arts
programs, 105–106
Intralinks (legal technology
company), 13, 31

Inventiveness/imagination, as
human traits, 3, 20–21, 64–65.
See also Creativity/creators;
Humans
Ireland, "National Skills Strategy
2025," 145–146

Jenkins, Henry, 160n15
Jennings, Ken, 77
Jeopardy (TV show), 77–78
Johnson, Lyndon B., 6
Johnson & Johnson, employment
with, 40
Jones, Mackenzie, co-op
experience, 97
Julian, David, 29–31

Kasparov, Gary, 19
Kensho (banking industry
software company), 30
Kerr, Clark, 136, 140
Keynes, John Maynard, 6
Knowledge economy
impact of machines on, xii, 26,
113
postwar development of, 11
and things left to learn and
discover, 47
Kolb, David, 83
KPMG consulting, employment
with, 40

Leadership skills, importance,
27